W9-CHK-887

WANDERING LANDS AND ANIMALS

BY THE SAME AUTHOR

DINOSAURS: *Their Discovery and Their World*

MEN AND DINOSAURS: *The Search in Field and Laboratory*

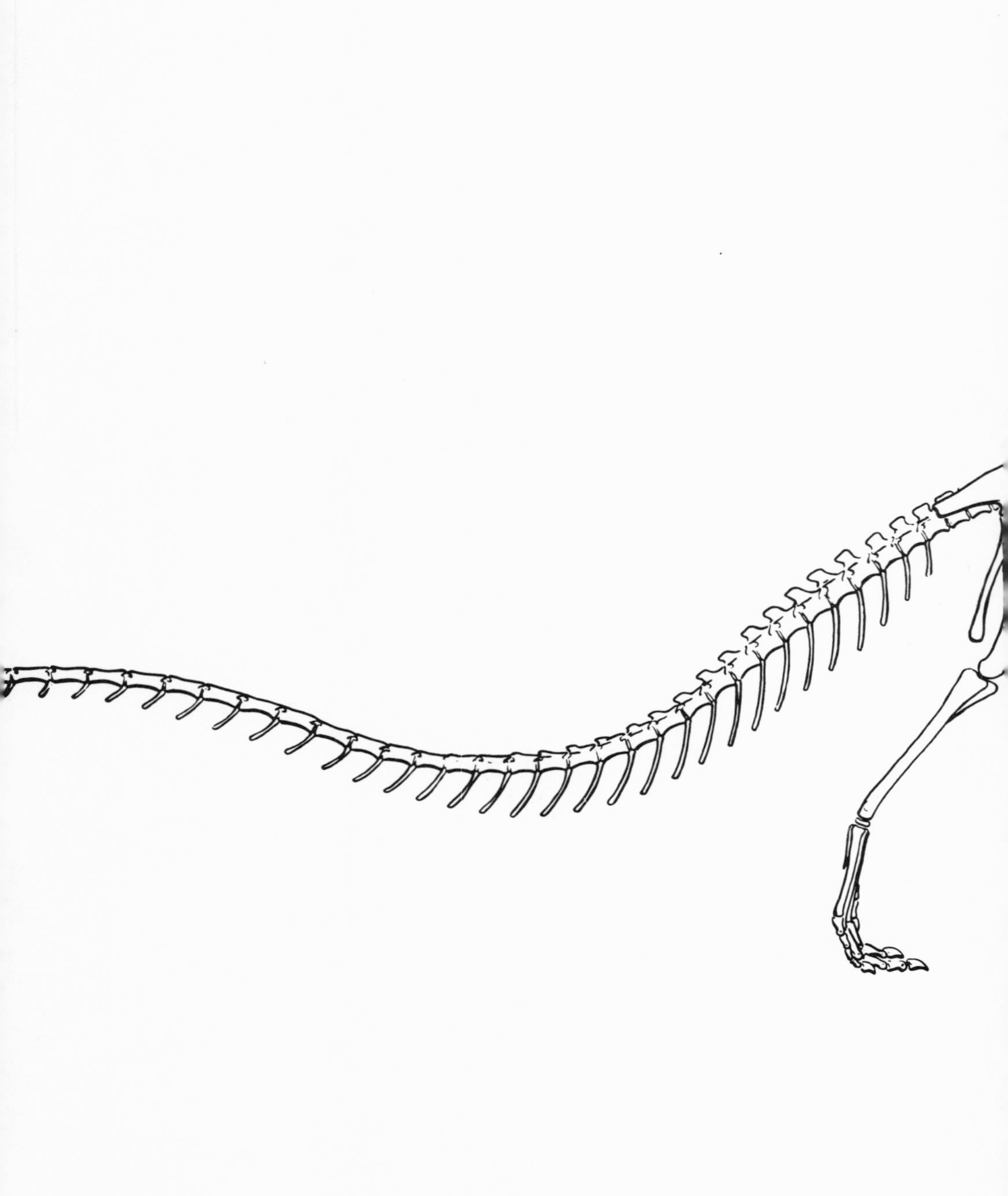

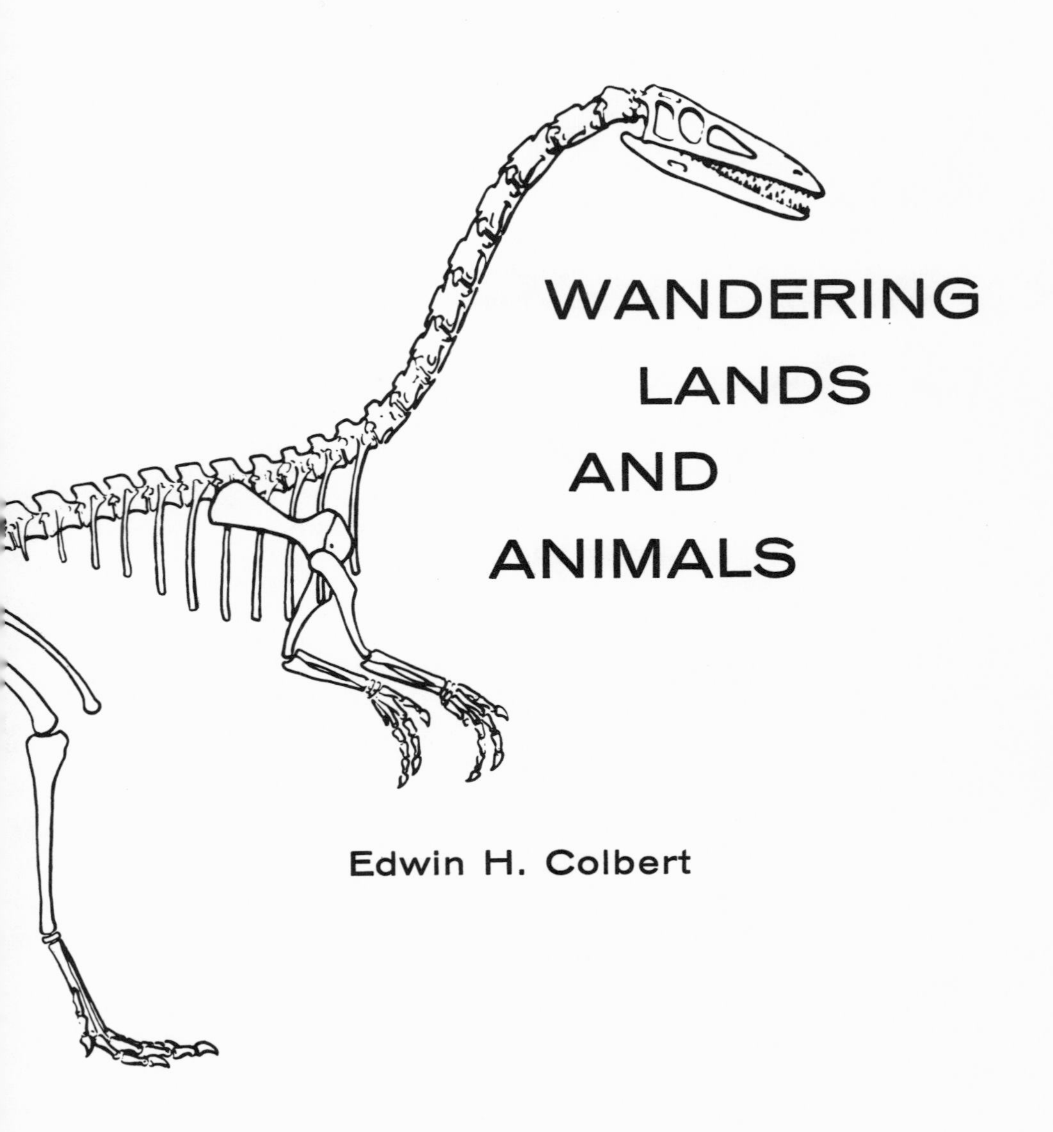

WANDERING LANDS AND ANIMALS

Edwin H. Colbert

E. P. DUTTON & COMPANY, INC.

NEW YORK · 1973

Grateful acknowledgment is made by the author for permission to reproduce the following copyrighted material.

FIGURE 3, from A. L. Du Toit, *Our Wandering Continents,* by permission of Oliver and Boyd.

FIGURE 4, from Alfred Wegener, *The Origins of Continents and Oceans,* by permission of Dover Publications, Inc.

FIGURE 33, from E. H. Colbert, *The Age of Reptiles,* by permission of W. W. Norton & Co., Inc., and George Weidenfeld & Nicolson Ltd.

FIGURE 37, from "The Oldest Tetrapods and Their Forerunners," by E. Jarvik, *Scientific Monthly,* Vol. 80, pp. 141–154, Fig. 11, March 1955, by permission of Erik Jarvik and *Science.* Copyright 1955 by the American Association for the Advancement of Science.

FIGURE 54, from E. H. Colbert, *Evolution of the Vertebrates,* by permission of John Wiley & Sons, Inc. Copyright © 1955, 1969, by John Wiley & Sons, Inc.

FIGURES 59, 79 and 89, from E. H. Colbert, *Evolution of the Vertebrates,* by permission of John Wiley & Sons, Inc., and the artist, Lois Darling. Copyright © 1955, 1969 by Lois Darling.

FIGURES 60 and 61, from F. von Huene, *Paläontologie und Phylogenie der Niederen Tetrapoden,* by permission of Gustav Fischer Verlag.

FIGURES 77 and 78, from Zofia Kielan-Jaworowska, *Hunting for Dinosaurs* (Cambridge, Mass., The MIT Press, 1969), by permission of the author.

FIGURES 85 and 88, from W. B. Scott, *A History of Land Mammals in the Western Hemisphere,* by permission of Mary B. Scott and Sarah P. Scott.

FIGURES 90, 96 and 98, from H. F. Osborn, *Men of the Old Stone Age,* by permission of Charles Scribner's Sons. Copyright 1915 by Charles Scribner's Sons.

FIGURES 91 and 92, from H. F. Osborn, *Proboscidea,* Vol. II, by permission of The American Museum of Natural History.

FIGURE 93, from Bjorn Kurtén, *Pleistocene Mammals of Europe* (London, George Weidenfeld and Nicolson Ltd., 1968, Chicago, Aldine Publishing Co., 1968). By permission of the author and George Weidenfeld and Nicolson Ltd.

FIGURE 97, from Marshall Kay and E. H. Colbert, *Stratigraphy and Life History,* by permission of John Wiley & Sons, Inc. Copyright © 1965, John Wiley & Sons, Inc.

Second Printing, April 1974

Published simultaneously in Canada by Clarke, Irwin & Company Limited, Toronto and Vancouver
SBN: 0-525-22976-0
Library of Congress Catalog Card Number: 76-158602
Designed by the Etheredges

CONTENTS

ILLUSTRATIONS

FOREWORD

By LAURENCE M. GOULD
Professor of Geology, University of Arizona
Former Chairman, Committee on Polar Research,
National Academy of Sciences

When Edwin H. Colbert and I were studying geology in our respective universities, the mass, the volume and the diameter of the earth were fixed, immutable inheritances. Furthermore, the tilt of the axis to the ecliptic had never changed. The North Magnetic Pole was north and the South was south and they had always been that way. The ocean basins and the continents were fixed, permanent features that had changed only in detail over the last six hundred millions of years at least. Our planet Earth itself was a dying body living on the inherited energy from its primal source—the sun.

We no longer accept any of these assumptions. Ours is not a dying planet. It is vigorous and dynamic. Creation continues.

This all comes about from the evidence for the theory of continental drift, or global plate tectonics, which has created the greatest revolution in man's thinking about his earth since the Copernican Revolution. Today geologists find themselves in much the same position that astronomers did at the time of Copernicus.

Today evidence from the varied disciplines of both geology and geophysics has validated the assumptions on which the theory of continental drift rests. This theory has simplified and unified all of the main fields in the earth sciences such as paleontology, paleobotany, sedimentary geology, ore deposits, seismology and volcanology.

Dr. Colbert examines the whole concept of continental drift through the eyes of a vertebrate paleontologist. This is a good approach, for similarities of fossils from widely separated areas led to the first coherent statement of the theory, that by Alfred Wegener in 1912. In examining the evidence for an original universal continent, Pangaea, and its breaking up into Laurasia, a northern hemispheric continent, and Gondwanaland in the southern hemisphere, detailed attention was first given to the latter. Here the evidence was more apparent in the nice fit of the coasts of Africa and South America and the similarity of both plant and animal fossils from widespread areas in southern land masses and India, with the exception of Antarctica. Wegener realized that if there had been such a single southern continent Antarctica must have been the heart or keystone of it. Indeed he cited Antarctica nineteen times in the 1915 edition of his book, *Die Entstehung der Kontinente und Ozeane,* and though a few fossils of the seed fern Glossopteris, so widely known from other southern lands, had been discovered early in the twentieth century, no animal fossils had yet been discovered and Wegener's theory received little support, especially in North America.

In spite of the great support and refinement given to Wegener's theory in 1937 by the South African geologist Du Toit in his book, *Our Wandering Continents* (one of the truly great classics in the history of geology), and the accumulating evidence of numerous invertebrate and plant fossils, there were many doubters even into the late 1960s. Birds might have carried plant seeds, invertebrate animals might have been rafted across the stormiest seas in the world on masses of vegetation or logs, reasoned the doubters.

Then in 1968 Peter Barrett of New Zealand brought back from Antarctica a fossil bone which Dr. Colbert identified as part of the lower jaw of a labyrinthodont amphibian.

This first vertebrate fossil from Antarctica was the promise of greater things to come and was sufficient to entice Dr. Colbert to join a field party from Ohio State University for the field season of 1969–70. And from an outcrop of early Triassic rocks some 300 miles from the South Pole Dr. Colbert did help find and did on December 4, himself identify a part of the jawbone of *Lystrosaurus,* a small reptile who lived over wide areas of other southern lands in early Mesozoic times some 200 million years ago. This was, of course, the very fossil Dr.

Colbert had gone all the way to Antarctica to find. It is altogether fitting that this should have happened. Darwin once observed that chance comes to the prepared mind. Edwin H. Colbert had that prepared mind. His highly productive life in the field of vertebrate paleontology had included extensive field work and collecting in the *Lystrosaurus*-rich rocks of India and South Africa.

A fellow geologist, Grover Murray, member of the National Science Board and I, as Chairman of the Committee on Polar Research of the National Academy of Sciences, arrived in Colbert's camp the day of this discovery. I still believe I did not exaggerate when I sent these words that night back to Washington: "This is the key index fossil of the lower Triassic in the major southern land masses of the former great southern continent of Gondwanaland. . . . . Not only the most important fossil ever found in Antarctica but one of the truly great fossil finds of all time."

Dr. Colbert's *Lystrosaurus* not only dated a continent, but it is also the keystone of the evidence for the reality of Gondwanaland. *Lystrosaurus* was a small, freshwater reptile tied to the earth who could not possibly have negotiated the wide seas that now separate southern land masses. No further proof was needed for the former existence of Gondwanaland.

Dr. Colbert's account of *Lystrosaurus* is the platform from which he takes off to recreate the whole magnificent panoply of land-living vertebrates, from the time of their emergence on land some 150 million years before *Lystrosaurus* down to and culminating in man.

Throughout this colorful and at times bizarre history, such as that of the dinosaurs, the relationship between the distribution of land vertebrates and continental land masses becomes increasingly clear. The terrestrial vertebrate fossil record alone establishes the validity of continental drift.

Dr. Colbert ends on a note of caution as he looks forward to the continuing evolution of the land vetebrates, because of the appearance of man. Man's assumption that the earth and its creatures were made for him, together with a mobility never known in geologic history, have already blurred the earlier reciprocal relationships between earth creatures and their environment. Dr. Colbert leaves open the question of whether man will learn before it is too late that he cannot bully nature but that he must cooperate and negotiate with it.

L. M. G.

PREFACE

This book presents a point of view—namely, that there has been through the immensity of geologic time a reasonably close relationship between the arrangement of the major land masses of the earth and the distributions of the animals inhabiting those lands. An understanding of why terrestrial animals differ on the several continents in which we live cannot be reached without a correlative understanding of the geographical relationships between the continents. And what holds for the present world holds in equal force for the world of the past.

Our understanding of the world of the past, however, depends upon the interpretation of a large amount of evidence that has been accumulated by many scientific disciplines. So it is that the interpretation of such evidence involves an exercise complementary to the relationship of continents and animal distributions, stated above. In other words, there is much to be gained in the reconstruction of relationships between continents of the past by the study of the distributions of extinct land-living animals as revealed by the locations of their fossil remains. To a large extent, such is the purpose of this book.

A major premise of the presentation made on the following pages is that an important clue to past continental relationships may be found in the relationships among fossil land-living animals—specifically fossil vertebrates—as they have been discovered throughout the world. In making this presentation the evidence has been restricted primarily to the earthbound animals with backbones, specifically the amphibians, reptiles, and mammals. For the most part these animals have of necessity moved from one place to another by dry-land routes, and for this reason they are of particular importance to the student who wishes to learn how lands were connected during past geologic ages.

Furthermore, this book is concerned with the past distributions of land-living vertebrates as they are related to the concept of the ancient supercontinents of Gondwanaland and Laurasia, and to the theory of continental drift. During the past two decades there has been an impressive accumulation of geological and geophysical evidence attesting to the probability of continental drift as the primary cause for the present arrangement of the continents—an arrangement derived from very different relationships of continents as they were 200 million years and more ago. It is here argued that the evidence of the fossil vertebrates is in accord with the concepts of an ancient Gondwanaland and Laurasia, of the fragmentation of these lands, and of the drifting of their several remnants to the positions occupied by the continents familiar to us. Perhaps the case might be stated the other way around: the ancient supercontinents and continental drift are in accord with the distributions of land-living vertebrates, past and present. The important point is that all of the evidence now seems to present an integrated picture of immense dimensions.

Perhaps the premise as presented herewith is wrong, but it certainly seems to make a lot of sense. The great picture that emerges from the discoveries of paleontology, biology, geology, geophysics, and other fields is far from perfect. But every year it seems to become increasingly clear, in large aspect and in detail. There is still much to be learned and much to be explained. And perhaps some of our present explanations will have to be revised. Nevertheless, one feels that the evidence as it accumulates consistently points to the probability that the world has been a mobile earth during the long years of geologic history and that this mobility is reflected in the rocks and the fossils of the earth's crust.

The writing of this book, hopefully intended for the general reader, has not been easy, because it attempts to deal with a large subject for which there is an incredibly vast literature, growing daily at such a rate that it is truly impossible to keep abreast of the never-ending stream of publications. Moreover, the literature on this subject embraces many fields of scientific endeavor, so that one person can hardly hope to have a true comprehension of more than a fraction of the whole. Consequently the book here presented is concerned primarily with continental vertebrates of the past and with the physical limits of the continents on which they lived. The overwhelming mass of geological and geophysical evidence that may be correlated with the subject at hand can be only briefly surveyed. There is all too much that must be omitted.

The author wishes to acknowledge the many sources from which the information for this book has been gathered. The works consulted are far too numerous to mention completely, but the more important and useful ones are listed in the bibliography. I have profited also from conversations with numerous paleontological colleagues, and I wish particularly to extend sincere thanks to Dr. David H. Elliot of The Ohio State University for critically reading sections of the manuscript dealing with the physical aspects of continental drift.

Errors of fact and of judgment must be the sole responsibility of the author.

The illustrations for the book were made by Mrs. Nova Young; the typing of the manuscript in its several manifestations was done by Mrs. Deloris Douglas. I wish to acknowledge my debt to the Museum of Northern Arizona in Flagstaff for facilities constantly available during work on the manuscript. Finally, I wish to express my appreciation to the publishers, and especially to Mr. William Doerflinger, editor, for help and encouragement.

EDWIN H. COLBERT

The Museum of Northern Arizona
Flagstaff, Arizona
December 1972

WANDERING LANDS AND ANIMALS

PROLOGUE

THE MODERN GEOLOGICAL REVOLUTION

Once upon a time the earth was thought to be flat. Not completely flat, because with rare exceptions there were mountains and hills and valleys to give variety to the landscape. Yet to the early inhabitants of the Mediterranean Basin, as was probably the case among people elsewhere, it was a flat earth arched by the dome of heaven, through which the stars winked at night and across which the sun followed his accustomed course from east to west during the day. It was all a matter of simple observation, and for untold years this view of the earth and the heavens above was sufficient. In the course of time, however, the Greek scholars came to a realization that the earth was not flat; rather it was a globe. Nevertheless, to all it was the center of the universe, circled by the sun and the moon, by the planets and the stars. Then there came the Copernican revolution in the early days of the sixteenth century, and the earthly globe was no longer the center of the universe. It was no longer a static earth, but a dynamic, whirling planet, and a small one at that, turning on its axis and circling a sun that was but one star in a great galaxy, which in turn was only one galaxy in a universe of galaxies stretching beyond the comprehension of the human mind.

But if the earth was no longer the center of the universe, Man was still at the middle of the biological world. All life was centered upon him. Then came the Darwinian revolution in 1859, and Man was no longer an exalted being, akin to the angels and far above the rest of the living world surrounding him. Moreover, life was no longer static and immutable on a static globe, but rather was dynamic and constantly changing through the course of time—time measured by the circling of the earth around the sun. Man's concept of nature and of his place in nature became a vision of energy received from the sun, of energy expended in living, and of constant change.

Thus modern man views the earth as an evolving, changing planet. And this vision has been especially apparent to the geologist—the student of the earth. Through two centuries successive generations of scholars devoted to the earth sciences have studied the rocks of the earth and the fossils contained within those rocks, and have reconstructed the events of earth history. They have seen the birth, the elevation and the destruction of mountain ranges, the varied development of rivers, the invasions of seas across lands and the emergence of continental areas above marine waters, and they have followed the evolution of plants and animals in this ever-changing scene. Yet until fairly recent years the students of the earth and of earth history have largely envisaged these changes, profound as they have been, as taking place on and around continents of comparatively fixed positions. It had long been supposed that the changes in the faces of the land have occurred on continents solidly anchored in the positions they now occupy.

Within the past few decades, however, this concept of static continents has been challenged. Geology is in the midst of a revolution, and new ideas, based upon evidence that was unforeseen a decade or two ago, are impinging upon the older concepts of our globe, to give us pictures of a past world at considerable variance with the views that had been delineated by earlier students of geology. It is a revolution more or less comparable to the revolution in biology that took place a century ago as a consequence of the publication of Darwin's epochal book *The Origin of Species*. But the geological revolution of today is, unlike the Darwinian revolution, less the work of one man than of many men, who have given their separate attentions to widely disparate lines of evidence. It is a revolution the consequences of which are being felt in ever-widening circles of scientific endeavor and will be felt with ever-increasing force as the years pass by.

In short, this geological revolution has involved the profound

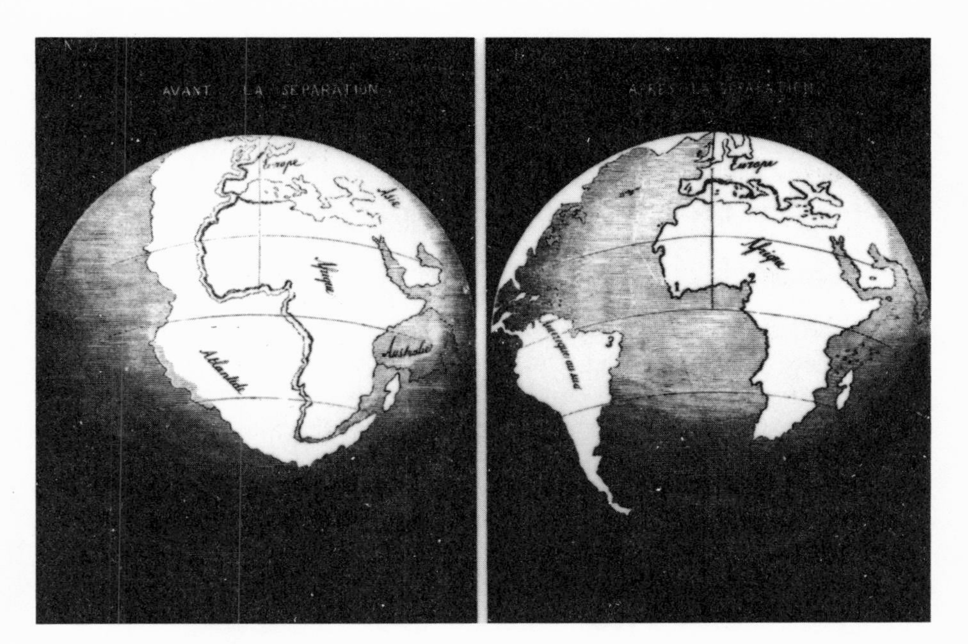

FIGURE 1. An early version of continental drift, stimulated by geological evidence. In 1858 Antonio Snider postulated an ancient contiguity of the continents on the two sides of the Atlantic Ocean to explain the presence of identical fossil plants in certain coal deposits of Europe and North America.

change of attitude resulting from modern evidence that would seem to favor the idea of continental drift. Have the continents been stable through the ages, occupying the positions on the globe that they occupy today, or have they been mobile, moving across the face of the earth through geologic time? Was the ancient world like the world of today so far as continental relationships are concerned, or was it vastly different from our modern world—so different as to be almost unrecognizable to one accustomed to the world in its present configuration? Was there once a single, immense southern continent (which has been christened "Gondwanaland"), composed of Africa, South America, Antarctica, Australia, New Zealand, and peninsular India, all intimately joined? Was there correlatively a single, immense northern continent (which has been christened "Laurasia"), composed of North America and Eurasia? Were these two continents joined to form a tremendous supercontinent (which has been christened "Pangaea")? Did the parent continents break asunder, their fragments drifting across the earth through time, to become the modern continents, shaped and positioned as we know them? These are questions centered upon the one problem that above all others, during these past 20 years or so, has engaged geologists and other people interested in the early history of the earth.

CONTINENTAL DRIFT—ORIGINS OF THE CONCEPT

The concept of continental drift is not a new idea. Indeed, its beginnings may go back as far as the days of Francis Bacon, to the time when the first reasonably accurate maps of the world revealed an astonishing parallelism between the outlines of the western border of the African continent and the eastern border of South America.* Certainly the significance of these continental outlines was receiving serious attention a century ago. And ever since those days this particular matter of continental shores on either side of the South Atlantic has attracted the attention of many people. In recent years it has stimulated renewed interest among numerous geologists. Can the correspondences in the South Atlantic coastlines of Africa and South America be entirely fortuitous?

These comparisons between the South Atlantic shores of Africa and South America bring us to the larger subject of Gondwanaland, already mentioned.†

Gondwanaland was a name proposed in the late nineteenth century by Eduard Suess, a famous Austrian geologist, to embrace a hypothetical continent of great geological antiquity. The name is derived from the land of the Gonds of India, a Dravidian people who established a kingdom in the central portion of the Indian peninsula, within the boundaries of which are exposed rocks that would seem to have belonged to this supposed ancient continent. Originally the concept of a Gondwanaland was introduced to include peninsular India, the island of Madagascar, and southern Africa, because of the similarities of rocks

* Bacon is widely credited in the literature on continental drift for having suggested that at one time Africa and South America were joined. But according to N. A. Rupke of Princeton University, no such proposition was made by Bacon. Rupke maintains that the first definite statement on this matter probably came from a German minister, Theodor Lilienthal, in 1756. According to Lilienthal, "the facing coasts of many countries, though separated by the sea, have a congruent shape, so that they would almost fill one another if they stood side by side; for example, the southern parts of America and Africa." (Calder, 1972, p. 42.)

† Of course, the concept of an immense southern-hemisphere continent, Gondwanaland, goes hand in hand with the correlative idea of an equally immense northern-hemisphere continent, Laurasia. At this place emphasis will be given to Gondwanaland, since it loomed large in the work of early proponents of continental drift.

and certain included fossils in these now separate regions. Subsequently, because of geological similarities not only between India, Madagascar, and Africa, but also South America and Australia, as well as Antarctica, the boundaries of this ancient continent of Gondwanaland came to include all of the now widely separate continents of the southern hemisphere, as well as the Indian peninsula. In the years following Suess's proposal many geologists, as well as biologists and other scientists, were favorable to the idea of an ancient Gondwanaland; but many were not. The concept was especially attractive to southern-hemisphere scientists, or to men who had done extensive geological and biological work in that part of the world. Clearly there were geological and biological resemblances between the land masses south of the equator, resemblances that crossed the equator to include a portion of India. Yet, in spite of such resemblances, many scientific workers, especially those of northern lands, were hostile to the idea. The theory of a stable earth, characterized by continental masses that had long been fixed in their present positions, was deeply entrenched in the literature and in the minds of geologists, as well as of other people. If the continents had been stably fixed through geologic ages, how was one to explain the resemblances between the several continents that theoretically once were parts of Gondwanaland?

One concept that persisted for decades envisaged a huge Gondwana continent, great portions of which broke down through extensive faulting and foundered into the depths of the southern oceans. Thus, according to this idea, Africa and South America and Australia, as well as peninsular India, are the remnants of a land mass that extended east and west across 270 degrees of latitude or more, most of this south of the equator. If so, the parts of Gondwanaland that disappeared beneath the seas would seem to have been as large as, perhaps larger than, the remnants represented by the modern southern continents. Many geologists objected, and rightly so, that this would have been a remarkably large amount of land to have disappeared within the ocean basins.

A variant of this theory has supposed that the Gondwana continents as we know them once connected by long, isthmian bridges, or perhaps by archipelagos. Such connections would have involved much less subsidence of land beneath the ocean waves than would have been required by the foundering of extensive continental areas. Even this was objected to by many authorities.

For many students of ancient life a continent of Gondwanaland

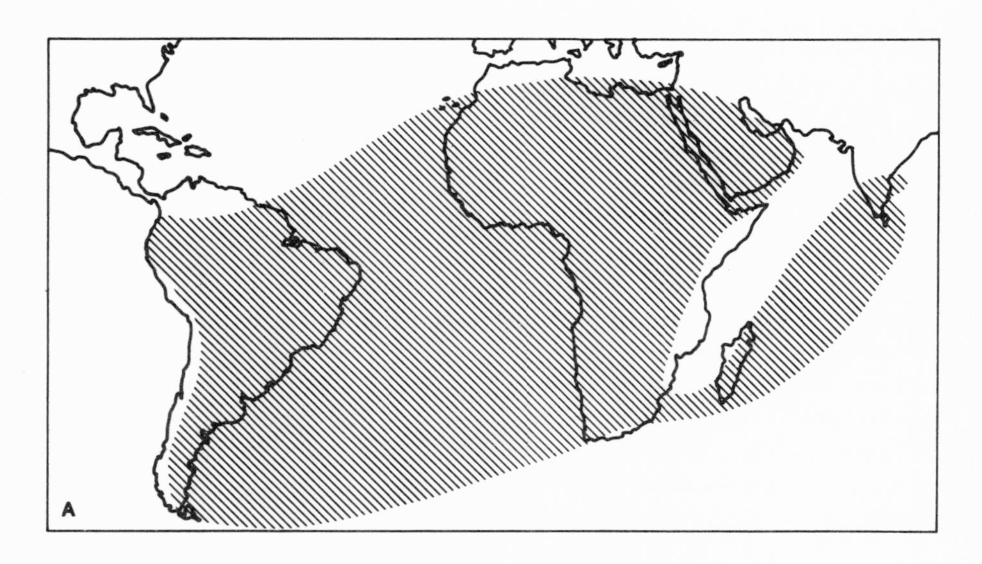

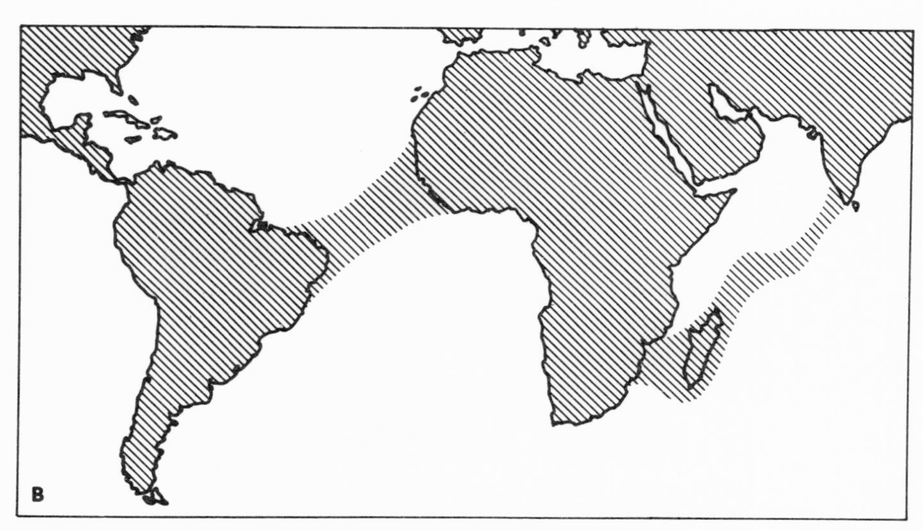

FIGURE 2. Two concepts of Gondwanaland, based on the theory of fixed continents. A. Gondwanaland (in part) pictured as a vast, east-to-west largely southern-hemisphere land mass (indicated by diagonal lines), of which the present southern continents are the remnants. According to this theory, large segments of Gondwanaland foundered into the ocean basins—a supposed geological event of enormous dimensions, for which there is no concrete evidence. (After Neumayr, 1887.) B. Gondwanaland pictured as composed of southern continents, connected by long, transoceanic land bridges, which subsequently sank beneath the waves. (After Schuchert, 1932.)

did not seem necessary. To explain the distributions of land-living animals through time, recourse might be had to the obvious modern paths for intercontinental migrations: along the Panamanian isthmus between the two Americas, across the Bering region (which during much of later geologic time, at least, was a land bridge), and through the East Indies to Australia (again over former land connections). Some of the more isolated land masses, such as New Zealand or Madagascar, were thought to have been populated by land-living animals rafted to these regions on floating logs or masses of vegetation. All such paths for intercontinental distributions often would have involved long, circuitous, and even hazardous journeys through space and time, but distributions of animals along these routes might not have been beyond the realm of possibility. It was also argued that many plants or even small invertebrate animals could have been distributed between the continents of the southern hemisphere by winds—in spite of strong

FIGURE 3. The so-called replacement globe as envisaged by H. B. Baker in a series of papers published from 1912 through 1914. Some aspects of this reassembly of Africa, South America, North America, and Europe resemble modern reconstructions of Pangaea.

objections on the part of some paleontologists and paleobotanists, the scientists who are concerned with ancient animals and plants, respectively.

While these conflicting ideas were occupying the thoughts of many geologists and biologists, a quite different theory was proposed to explain the existence of a former Gondwanaland. This was a modern expansion of the old observation that the opposing borders of Africa and South America can be fitted against each other. In 1912 a German scientist, Alfred Wegener, published an epochal book, *Die Entstehung der Kontinente,* in which (as translated) he stated that "he who examines the opposite coasts of the South Atlantic Ocean must be somewhat struck by the similarity of the shapes of the coastlines of Brazil and Africa. Not only does the great right-angled bend formed by the Brazilian coast at Cape San Roque find its exact counterpart in the reentrant angle of the African coastline near the Cameroons, but also, south of these two corresponding points, every projection on the Brazilian side corresponds to a similar shaped bay in the African." As we have seen, this was not a particularly new observation, but in the mind of Wegener it led to an idea of great import—the theory of continental drift. The nub of the idea is that our continents were not fixed through geological time, but rather that they drifted across the face of the globe to their present positions. According to this theory, the southern-hemisphere continents and India might be the displaced remnants of a great Gondwana continent that broke apart at some time in the geologic past. Although Wegener was a pioneer in the development of the theory of drifting continents, he was not alone, for at essentially the same time that the concept of drift was envisaged by him, an American geologist, F. B. Taylor, also suggested the possibility of continental drift. Coincidences such as this are not uncommon in scientific endeavor; their occurrence is probably owing to the fact that the scientific climate, the result of the accumulation of facts, is propitious for the development of a theory to explain those facts. Thus Darwin and Wallace in 1858 proposed their epochal theory of organic evolution together, and thus Wegener and Taylor between 1910 and 1912 came to the conclusion that continents in earlier times might have drifted across the surface of the earth. But even Wegener and Taylor were not the first to suppose such movements of continental masses. As long ago as 1857 an author named W. L. Green, writing in the *Edinburgh New Philosophical Journal,* spoke of "segments of the earth's crust which float on the liquid core." And in 1880 H. Wettstein pub-

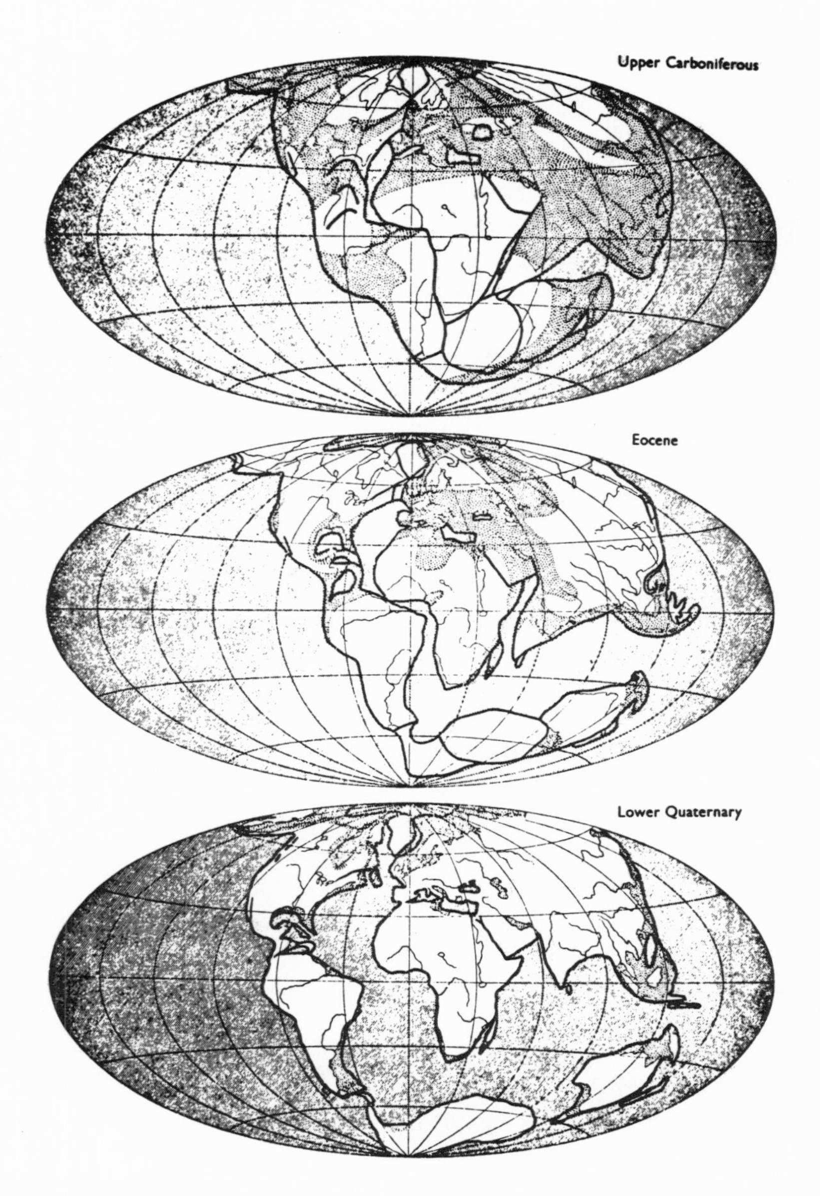

FIGURE 4. A famous illustration. Wegener's reconstruction of a map of the world according to his drift theory, for three stages of geologic history. These maps, based on the relatively meager evidence available to Wegener, anticipate to a remarkable degree the modern reconstructions, based upon sophisticated geophysical, geological, and paleontological researches.

lished a book in which he suggested that there had been horizontal movements of the continents—an idea that was echoed by various other writers, some qualified, others not, between then and the time when Wegener and Taylor first published their conclusions.

Although Taylor's concept of continental drift, published in 1910, slightly antedated that of Wegener, published in 1912, the credit properly can go to Wegener, for he is the one who truly developed the idea of continental drift, just as it was Darwin who developed in detail the theory of evolution. Darwin had his champion in the person of Thomas Henry Huxley, and Wegener had his champion in the South African geologist A. L. Du Toit. Indeed, Du Toit developed the theory of continental drift far beyond the limits that had been reached by Wegener, so that in many respects our modern ideas concerning drift are heavily dependent upon the work of Du Toit. Du Toit was a southern-hemisphere geologist, whereas Wegener was a German; it is probable that Wegener, in spite of his originality of thought, lacked the insight and certainly the familiarity with southern-hemisphere geology, which was the strength behind the work of Du Toit. Du Toit made many perceptive conclusions; from his intimate knowledge of southern-hemisphere geology he had a magnificent vision of ancient Gondwanaland, of its rupture, and of the drifting of its fragments to their present positions in the southern hemisphere and at the base of the Asiatic continent. But Du Toit lacked many of the hard geologic facts to give true substance to his vision, and for this he was severely criticized. Now it is fascinating, at this late date, to see how modern, solidly based work in varied fields of geological research has corroborated the ideas that had been reached several decades ago by Du Toit. Today we have the advantage of hindsight based upon new knowledge —knowledge of a sort that was not even guessed at by Du Toit and his contemporaries.

The former resistance of numerous geological authorities to the idea of continental drift was based not only on the tenuous evidence then available for support of the theory, but also to the almost impassioned arguments that frequently were put forward by its supporters. In short, the proponents of continental drift and Gondwanaland often stretched the evidence beyond the limits of what then seemed to be credibility, thereby damaging what otherwise might have been good arguments in favor of the theory. Theirs often seemed to be a sort of special pleading, viewed by their antagonists as being something less than objective in foundation and expression. So it was that many sober, sincere geologists looked down their collective noses at the whole business of Gondwanaland, of drift, and of drifters.

This writer recalls a Christmas party, perhaps 20 years ago, in a university geology department of outstanding reputation, at which one of the professors, a man of worldwide fame, a leader in his field, and an authority strongly opposed to the theory of continental drift, was presented with a large wooden anchor by his graduate students, to be used for securing the continents against drifting. There was much hilarity, and a good time was had by all—but just the same the humor pointed up a serious and basic tenet at the time. Continental drift was just not quite respectable in North America.

CONTINENTAL DRIFT—ITS MODERN REVIVAL

That was immediately preceding the widespread and detailed geophysical and oceanographic work of the past two decades that has produced our modern knowledge of the bottom of the sea, of great ridges running through the lengths of midoceanic regions, of profoundly deep oceanic trenches bordering various continental blocks, of the configurations of the edges of those blocks, of paleomagnetic measurements that indicate the lateral spread of the sea floor away from the midoceanic ridges, and of other paleomagnetic studies that show the former positions of the earthly poles. As a result of such studies much evidence has been brought to bear on the problem of the former positions and relationships of the continents, with consequent new and fresh views concerning this great and fundamental geological problem.

All of which has brought about the vigorous revival and extension of the theory of continental drift. New and often exciting evidence would seem to point in increasingly strong terms to the probability of drifting continents. Moreover, the many lines of new evidence within various, widely differing fields of scientific endeavor appear with increasing frequency to be correlated, meshing together to create a unified large concept, within which the lesser components are in harmony.

A theory, to be valid, must satisfy all aspects of the subject upon which it touches. So far as the theory of continental drift is concerned, many paleontologists have to a large degree been in something of a dilemma. For, as we have seen, to these men the drifting apart of our continents has not seemed to be of vital necessity in order to explain the relationships and distributions of former life on the earth. To them

FIGURE 5. The first fossil of a Triassic backboned animal to be discovered in Antarctica. A jaw fragment of a labyrinthodont amphibian, as it was seen by Peter Barrett—imbedded in coarse sandstones of the Fremouw Formation, high on the slopes of Graphite Peak, about 400 miles from the South Pole. Quite obviously it took a sharp and discerning eye to see this epoch-making fossil in the rock.

it has often seemed that continental drift raised as many problems as it solved.

CONTINENTAL DRIFT AND PALEONTOLOGISTS

Here there has been something of a geographical division, between paleontologists of the northern and southern hemispheres, and to a lesser extent, between those of the eastern and western hemispheres. The division of opinion can also be equated to a considerable extent with the age of the rocks and fossils with which the paleontologists have been concerned. Men working on the older rocks and fossils of the earth's crust have in numerous instances been more favorable to continental drift than those whose interests have been centered upon the younger rocks and fossils. Therefore, while some paleontologists have been converted to drift, others have been inclined to view it still with some degree of skepticism. These latter have seen means other than the direct joining of continents for land-living animals to move from one land mass to another.

This conservative view received something of a jolt in January 1968, when Peter Barrett, a New Zealand geologist at that time associated with the Institute of Polar Studies at The Ohio State University, returned from Antarctica bearing a small fragment of fossil bone. He had found the fossil high on the side of a mountain known as Graphite Peak, some miles to the east of the great Beardmore Glacier and within about 400 miles of the South Pole. The fossil (it was submitted to me for identification) proved to be a portion of the lower jaw of a labyrinthodont amphibian. It had been found in place, in rocks of early Triassic age, which would have made it something on the order of 200 million years old.

FIGURE 6. The jaw fragment, shown in figure 5, after its removal from the rock. This is a portion of the left jaw, in the region of its articulation with the skull. The rather semicircular notch, seen on the upper border of the jaw, is the articular surface that impinged upon the knoblike quadrate bone of the skull.

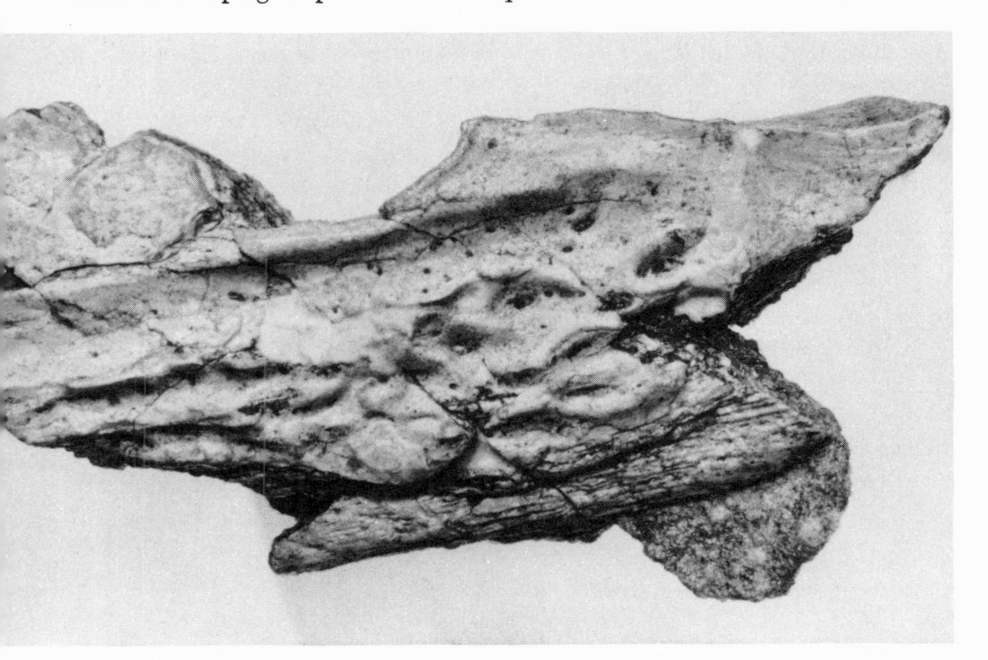

AN ANTARCTIC FOSSIL

It was an exciting and significant discovery, and it created a bit of a sensation, not only in geological and paleontological circles, but among the general public as well. For this fossil, esoteric though it might be, could be appreciated, and its significance evaluated, by the readers of newspapers, by people with no paleontological training. Here was a fossilized bone, broken and imperfect, yet well preserved, of a land-living animal on a now completely isolated continent. Moreover, here was a fragment representing an animal that must have lived in a benign, if not a tropical climate, now locked in the ice-bound rocks of the Transantarctic mountains, far within the coldest and most inimical continent on the earth. What did this fossil mean? How had the animal that it represented reached the locality where the bone was found?

It meant that, unless this fossil occurrence was a complete fluke of some sort, land-living vertebrates had reached Antarctica by an overland route at the beginning of Triassic times, more than 200 million years ago. It was significant because it pointed to great possibilities: where there was one such fossil logically there should be more. It meant the opening of new vistas in the study of continental relationships as revealed by remains of the ancient animals that dwelt upon former continents.

Since I had become involved in the identification of the single jaw fragment from Graphite Peak, inevitably I became involved in a detailed search for Antarctic Triassic tetrapods (the tetrapods being four-footed animals with backbones, in this case ancient amphibians and reptiles). The single fragment of amphibian jaw was convincing to me; and I then joined the ranks of the drifters—toward which I had been strongly inclined for several years. A field party was organized that went to Antarctica, to the Transantarctic mountains in the region of the Beardmore Glacier, during the austral summer of 1969 and 1970, hopefully to find fossils. Our efforts were eminently successful. Consequently another party went back to the Transantarctics in the season of 1970–1971, to bring back more fossils, many of them more complete and more widely representative of ancient Triassic life in Antarctica than were the fossils of the previous year.

These fossils, and their bearing upon the problem of continental drift, is in part what this book is about. But only in part—for the problems of drifting continents and the relationships of land-living animals extend far beyond the bounds of Triassic amphibians and reptiles in Antarctica. The land-living backboned animals that inhabited Antarctica when it was a tropical or subtropical land covered with ferns and forests afford us a point of departure, from which wide explorations into the past may be made. These explorations as set forth in the following pages will furnish, it is to be hoped, the information for a series of delineations showing where land-living vertebrates lived during past geological ages and how their lives were affected by the changing positions of the continents through time.

It is a large subject.

I. LYSTROSAURUS

THE CLIFFS OF ANTARCTICA

The Antarctic sun was bright in the clear, blue sky, continuing its seemingly never-ending circle high above the horizon—from east to north to west to south—flooding the empty land with its cheerful light. The warmth of the sunlight felt good on one's face. Far and away, across countless miles of white landscape, the almost limitless ice fields glittered beneath the brilliant sun, and the cliffs of the Transantarctic mountains formed dark ramparts against the blinding reflections from the vast expanses of snow and ice. It was a mild summer day in Antarctica, and yet there was a bite in the air. A sharp wind blew in from the south, flowing down over the plateau from the pole, and this wind countered some of the illusion of summer. It was cold against one's face, so that the tendency was to turn one's back, to let the wind beat against the padded insulation of one's parka. In that way it was possible to be reasonably comfortable.

We were working along the cliffs of Coalsack Bluff ("we" being William Breed, James Jensen, Jon Powell, and myself), where we had been working for a week or more, searching for the fossil bones of long-extinct animals and chiseling them out of the coarse sandstone. The

FIGURE 7. Off for a day's work at Coalsack Bluff, seen in the background. The fossil exposures are on the far side of the "bluff," more properly a nunatak (the isolated tip of a mountain, the bulk of which is buried by ice), which drops in a series of cliffs and long slopes to a lower ice field.

cliffs formed a broken escarpment that, for the better part of a mile, interrupted the long, steep slope of the bluff. Generally these cliffs were 20 or 30 feet high, and in most places it was not a very difficult feat to scramble up and down their faces in search of fossils; a welcome relief from the dogged, tiring climb along the loose, slanting slopes of the bluff that was necessary to reach these cliffs. Above the cliffs the slope continued to more low cliffs, and above them to a jagged wall of hard volcanic rocks. This high palisade of ancient magmas, which once had welled up from the interior of the earth, durable though it might be, had nevertheless been in part broken and fractured by the inexorable forces of wind and ice, so that its eroded remnants, consisting of countless slabs of glistening, wind-polished rocks, covered the slope like a loose pavement. The slope, with its pavement of volcanic slabs, continued below the fossil cliffs, as well as above them, down and down to the foot of the bluff, where it met the ice field along a part of its margin, and where, along another part of its margin, it terminated against the border of a tremendous moraine, the surface of which was a jumble of large boulders.

FIGURE 8. The friendly Antarctic. Looking out from the fossil exposure of Coalsack Bluff, across a vast panorama of snow, ice and rocks.

The fossiliferous cliffs, the focus of our attention, were in truth a small part of the bluff, just as the bluff itself was a small feature in the magnificently grand sweep of Antarctic scenery. The sandstones that formed the cliffs were irregular in shape; they had been formed by swift-flowing streams that had swept sands along to deposit them in sloping bars at the bottoms and along the sides of the streambeds. Thus, as revealed in the cliffs, the sandstones were seen as intersecting festoons of glittering grains, and within these festooned sands, known as "cross-beds" to the geologist, were enclosed the scattered bones of ancient reptiles and amphibians, the former predecessors, far distant in time, of the dinosaurs and more familiarly of our modern crocodiles, turtles, lizards, and snakes, the latter likewise long-extinct predecessors of our modern frogs and salamanders. Those bones were the objects of our search.

Coalsack Bluff, large enough in human terms, was a very small feature in the immensity of Antarctica, just as the fossiliferous cliffs were in turn a small part of Coalsack Bluff. Furthermore, the bones that we were collecting were small fossils, many of them no larger than a fingernail, none of them more than a few inches in length.

FIGURE 9. Exploring the Triassic sandstones at Coalsack Bluff for fossils. W. J. Breed in the foreground.

Nevertheless, they were extraordinarily important fossils, because they were, with the exception of the single fragment of jaw found two years previously, the first ancient bones from the Antarctic continent. As such, they received our very careful attention.

We had already collected more than a hundred fossil bones, some of them nicely preserved and some of them mere fragments. But as yet we had no very exact clues as to what the bones might represent. True enough, they were undoubted amphibian and reptile bones. Moreover, the amphibian bones quite definitely represented a group of ancient amphibians known as "labyrinthodonts"—the amphibians that had ruled the continents briefly some 300 million years ago and which for many millions of years thereafter had shared these continents with the early reptiles. Also, the reptile bones were obviously in part representative of ancient but progressive reptiles known as "therapsids," or mammallike reptiles, and in part representative of forms contemporary with the later therapsids, reptiles known as "thecodonts," these including the ancestors of the crocodiles, the flying reptiles, the dinosaurs, and also the birds, all of which were to dominate the world in later geological ages. But although the general nature of our fossils was thus apparent, we still were unable to be sure of the detailed, close identification of any particular bone.

COALSACK BLUFF AND GEOLOGIC TIME

There was good reason to think that our fossils were coming from rocks of early Triassic age, from the geologic period that marked the beginning of the Age of Dinosaurs. The association of labyrinthodont amphibians with therapsid reptiles and with thecodonts pointed to this, although such an association might conceivably indicate a later phase in Triassic history as well. However, various other aspects of the geology of the area indicated that the rocks and their contained fossils were probably within the limits of the Lower Triassic. For instance, the lowest part of Coalsack Bluff consists of dark shales with included coal seams, together designated as the Buckley Formation, and within many of the shales are the abundant fossilized leaves of a plant known as "*Glossopteris*." This plant is very characteristic of Permian rocks in the southern hemisphere. Immediately above the *Glossopteris*-bearing Buckley shales are sandstones, included within the recently named Fremouw Formation, among which are the cliffs containing the fossil bones. The Fremouw sandstones, following the Permian shales in close succession, would lead one to think that they are probably of Lower Triassic affinities. Moreover, the presence of such fresh-water sediments, their upper surfaces covered or capped and their layers invaded or intruded by extensive volcanic rocks, which is so typical of Coalsack Bluff, and of the Transantarctic mountains in general, is also very characteristic of the Permian and Triassic geology of South Africa. All in all, everything seemed to point to an early Triassic age for our fossils.

To say that the fossil bones at Coalsack Bluff are of early Triassic age is to speak in geological terms, clear enough to those who have had an introduction to the earth sciences, but perhaps of small significance to the person for whom geology is as yet an unopened book. But to say that the fossil bones at Coalsack Bluff are on the order of about 200 million years old is to identify their place in earth history in generally familiar terms. These fossils are old, yet many fossils are much older, just as many are much younger than the bones of Coalsack Bluff. And the earth is far more ancient than the oldest fossils known to us.

According to the best radiometric measurements of the rocks of the earth's crust, it is now evident that our planet was born at least four billion years ago. Although there must have been life on the earth at a very ancient time, the first adequate record of life, as revealed by fossils, is found in rocks about 600 million years old. From that point on to a geological yesterday, the fossil record continues in ever-increasing complexity, revealing to the eye of the paleontologist a story of progressive evolution from simple plants through primitive plants to the marvelous array of vegetation that covers most of the lands of the earth today, and from early animals without backbones to increasingly complex and numerous members of the invertebrate phyla, to primitive fishes, and on through cold-blooded amphibians and reptiles to the warm-blooded birds and mammals that inhabit the modern world.

This petrified record of life through the ages has been divided into three great eras of earth history, each of which in turn is further subdivided into several periods and lesser categories of time. The three eras of life are the Paleozoic—the era of ancient life, when at first invertebrates and primitive plants inhabited the earth but when, by the end of the era, plants were much advanced, and backboned animals in the form of varied fishes, amphibians and many reptiles lived in profusion around the circumference of the globe; the Mesozoic—the era of middle life, when for much of this time coniferous trees were ubiquitous and dinosaurs ruled the earth, but when, at the close of the era, modern flowering plants appeared, as did birds and primitive mammals; and the Cenozoic—the era of modern life, when the plant life assumed aspects familiar to us, and when modern fishes evolved in all of their abundance and complexity, while birds and mammals ruled the continents. At the end of the Cenozoic era man appeared.

The record of life on the earth may be presented in the form of a table, as is common in textbooks of geology, with the oldest time divisions at the bottom of the chart and the youngest at the top (because the fossiliferous rocks of the earth's crust quite obviously have been deposited with the oldest rocks and their contained fossils at the bottom with successively younger rocks and fossils following upward in sequence).

So it can be seen that the early Triassic fossils we were finding, even though immensely ancient from our point in the perspective of time, were still considerably less than half as old as the fossil record. To us they were important, for they came from a crucial period of earth history—a time when the ancient life of this planet was giving way to evolutionary pressures exerted by new and vigorous plants and animals, organisms that were to dominate the earth for the next 100 million years and more. Moreover, it would seem probable that this Triassic period, with its influx of exuberant animals and plants, was also a time crucial in the crustal history of the earth. Consequently, we were more than anxious to learn something about the exact nature of the fossils we were finding—in order that we might gain some clue as to their bearing on the development of life in the southern hemisphere and on the role of Antarctica in the evolution of earth history.

If only we could discover some bones—even one bone—that would give us a definite lead here in the field as to the nature of the animal it represented. Of course, once we got back to the United States, with resources of museums and their collections and libraries with their literature on Triassic fossils, it would be possible to run down most, if not all, of the fossils in our growing collection. But in the field, without

ERAS	PERIODS AND EPOCHS		DURATION, m.y.[1]	LIFE
Cenozoic 65 m.y.	Quaternary	Pleistocene	3	Man supreme; Appearance of Man
	Tertiary	Pliocene	9	Triumph of mammals, birds
		Miocene	15	Bony fishes
		Oligocene	10	Rise of modern mammals; Modern invertebrates
		Eocene	18	Flowering plants
		Paleocene	10	Primitive mammals
Mesozoic 160 m.y.	Cretaceous		70	Ancestral mammals; Advent of flowering plants; Dinosaurs supreme
	Jurassic		60	Giant dinosaurs; First birds
	Triassic		30	First dinosaurs; "Ganoid" fishes; Conifers
Paleozoic 375 m.y.	Permian		55	Early reptiles
	Carboniferous[2]		65	Large amphibians; Ancient plants
	Devonian		50	Primitive fishes; Primitive plants
	Silurian		45	First backboned animals
	Ordovician		60	Invertebrates supreme
	Cambrian		100	

Precambrian time—about 3½ billion years.

[1] m.y. = millions of years.

[2] American geologists usually recognize two periods in place of the Carboniferous: an earlier Mississippian period and a later Pennsylvania period.

FIGURE 10. Collecting fossils at Coalsack Bluff. In the foreground James A. Jensen is hard at work excavating a skull fragment of *Lystrosaurus*—the first truly recognizable fossil of this reptile to be found in Antarctica. The date was December 4, 1969, and it was a cold, unpleasant day to be out in the field. (W. J. Breed in the background.)

comparative collections and books at hand, identifications were in many ways of the "by guess and by gosh" sort, and only truly distinctive fossils could give us the clues we needed.

So on that bright but windy day out on the cliffs and the loose, rock-covered slopes of Coalsack Bluff, I pondered the problem as we chiseled away at the fossil bones. And I thought to myself, "Now, why can't we find a good bone of *Lystrosaurus?*"

Why *Lystrosaurus?*

LYSTROSAURUS IN ANTARCTICA

Because *Lystrosaurus*, typical of the Lower Triassic of Africa and India, is one of those very useful fossils often known as "guide fossils." It is of distinctive aspect, so that it can be readily recognized, especially if one has the right parts; it is stratigraphically confined, which means that it does not extend vertically through a wide range of sediments (in other words, its geologic time span was relatively short); and it is frequently very abundant in the rocks, so that it is easily discovered. In fact, the lowest beds of Triassic rocks in South Africa, some 220 million years in age, are known as the "*Lystrosaurus* zone," so abundant and characteristic is this fossil in the sediments.

In this therapsid, or mammallike, reptile the bones of the skeleton are, of course, different from the bones of other therapsids, but the differences are often small and of a subtle nature. In the laboratory they may be detected; in the field they may not be so apparent unless one has been doing a lot of detailed work on *Lystrosaurus* and is thus familiar with the minutiae of its anatomy. In form, this reptile may be up to four or five feet in length, with a rather heavy skeleton, a barrel-like body, stout, stocky legs, broad feet, and a short tail.

If there is nothing especially noteworthy about the skeleton, the same cannot be said of the skull and jaws, for in *Lystrosaurus* the head was remarkably specialized. The front of the skull, in front of the eye, is strongly downturned, so that the nasal region is at about right angles to the top of the skull. The nostrils are in a rather high position, in front of the eyes. The lower jaw is rather like the beak of a turtle, and bites against a broad, beaklike upper jaw. Of especial significance is

FIGURE 11. A fossil *in situ* at Coalsack Bluff. David Elliott (left) and E. H. Colbert are quite pleased to look at a *Lystrosaurus* bone, before it is chiseled out of the rock.

FIGURE 12. A reconstruction of *Lystrosaurus.* This interesting and important reptile was a heavily built animal, with a rather barrellike body, short, strong legs and broad feet, and with a most distinctive head, characterized by the beaklike jaws and the pair of tusks in the skull.

the fact that there are no teeth—hence the beaklike jaws—except for two large tusks, one on each side, in the skull. These characteristics give to the skull of *Lystrosaurus* a distinctive aspect that can hardly be mistaken.

By virtue of the two tusks in the skull, *Lystrosaurus* is a member of that group of therapsid, or mammallike, reptiles known as "dicynodonts" (*di*—two, *cynos*—dog or doglike, *odontos*—teeth). The dicynodonts are extremely numerous and varied in the Permian and Triassic sediments of various continents, especially Africa, and for millions of years, through a span of some 50 million years, from about 250 to 200 million years ago, they must have been extraordinarily successful reptiles.

There has been much speculation as to the evolutionary significance of the strangely shaped skull and the robust body of *Lystrosaurus.* These characters, which make of *Lystrosaurus* such a strange-looking reptile (in our eyes), would seem to indicate that it was probably an inhabitant of streams, rivers, and lakes; the high position of the nostrils and the barrellike body point to this supposition. Moreover, the remains of *Lystrosaurus* are found in Africa in sediments that clearly were deposited in the beds of streams and lakes. At Coalsack Bluff, as well, the cross-bedded, festooned sands limn an unequivocal picture to the geologist; they trace the downward grade of ancient streambeds.

But Coalsack Bluff today is far removed in time and in environment from the world of *Lystrosaurus*. This ancient reptile must have been an animal of the tropics, or at least the subtropics, inhabiting a well-watered region of abundant plant growth. There is every reason to think that *Lystrosaurus* in Africa was a herbivore. Its abundance in the early Triassic sediments of that continent would indicate that it lived in great numbers—in herds, if you wish to use that term—as do various modern plant-eating animals. Certainly the structure of the jaws would be well-adapted for cutting and nipping green plants, and certainly the heavy, barrellike body gives circumstantial evidence that this strange reptile lived on a bulky plant diet. Thus one may picture *Lystrosaurus* as living a life rather similar to that of a modern hippopotamus (but on a reduced scale and some 200 million years removed in time). It is an exotic picture to conjure up on the icy cliffs of the Antarctic.

So, if the festooned sands of the Fremouw Formation are of early Triassic age, perhaps they should contain *Lystrosaurus*. And *Lystrosaurus*, if present, would determine the age and the geographic relationships of these rocks near the South Pole, as almost no other fossil could. *Lystrosaurus* would certainly indicate the close relationship of Antarctica to Africa, in unequivocal terms. And that is why, out on the windy slope of Coalsack Bluff, I wistfully hoped that perhaps a definitive bone of *Lystrosaurus* would be revealed.

The afternoon wore on, and at length it was time for us to return to camp. We packed our tools and such fossils as had been collected during the day into haversacks, and began the slow and tiring trek along the face of the bluff to the head of a col, where we had parked our motor toboggan and sled. Climbing along Coalsack Bluff was one of the most tedious and wearing aspects of our work there. The interior of Antarctica, being devoid of life, is a land of snow and ice, bare rock faces and loose talus slopes. There is no vegetation to hold the dirt, so that on those slopes lacking snow the loose soil, if so it may be called, rests at the natural angle of repose, which is pretty steep. And at Coalsack Bluff this slope is covered by the loose pavement of weathered volcanic slabs. Consequently a hike along the face of the bluff was an exercise in vigorous scrambling punctuated by a modicum of profanity, in an effort to move horizontally. At every step one would slide down the slope, so it was a constant fight uphill to maintain a horizontal course. Therefore, we usually arrived at our sleds in a blown condition and frequently very much heated up inside the insulation of our heavy Antarctic field clothing.

But at last we reached our sleds, and from there into camp we had an easy and at times exciting ride. It was a long downhill journey across the ice, for the camp was many hundred feet below the level of the bluff. The motor sled would chug away at a steady pace of about ten miles per hour, but the Nansen sled, the standard light sled used for decades by polar explorers—a beautifully designed wooden sled, its various members lashed together with leather thongs, thus making it flexible, and remarkably suited for rough ice and snow surfaces—which theoretically was being towed, had a tendency to run ahead of the power sled down the slopes. Since the motor sled would carry only one or at the most two people, two or three of us perforce rode as passengers on the Nansen sled. For us the ride had its thrilling moments, as we scooted past the motor sled and then skidded sideways at the end of our towing rope, like the end members of a crack-the-whip game.

It was good to be back in the warmth and comfort of camp after a day on the bluff. And here a few words might be said about our camp. It was no simple affair, but rather a carefully planned and comparatively large establishment. The four of us who were collecting the bones of Triassic animals were part of a larger group of about 20 geologists and paleontologists making a detailed study of the Transantarctic mountains. The project had been planned by the Institute of Polar Studies, which is located at The Ohio State University in Columbus, and had recruited a rather cosmopolitan group of men, including some from foreign countries, to make this survey. All of this work was under the auspices of the National Science Foundation of the United States, and since the large perennial scientific effort in the Antarctic is a function of a governmental agency, the logistics of the program—and they are large and complex—are provided by the Navy. Only with such great resources at hand is it possible to carry on scientific work in Antarctica efficiently. Only by having ample and frequently expensive support available is it possible for scientists to devote attention to their problems, rather than to have to spend a major portion of their energies

FIGURE 13. Camp at Coalsack Bluff. The Jamesway huts give appreciated protection and warmth against the howling winds that blow down from the south polar plateau.

FIGURE 14. Equipment for the south polar paleontologist. Here one sees two motor sleds, the front of a Nansen sled, the "crabhut"—a specially designed tent for use on fossil exposure—skis, and more mundane objects.

on the business of keeping alive, as was the case in the early days of Antarctic exploration.

So it was that we had a large camp, housing not only the score of scientists who were working on many phases of Transantarctic geology, but also an equal number of Navy personnel for support. There were several pilots in our camp, because we had some helicopters available for transport to geological sites that were far removed from our base, and as a complement to the pilots there was a crew of mechanics. And, of course, a cook and his helper, a medical corpsman, radio operators, and others. Thus we were a sizable and busy community to be living within 400 miles of the South Pole.

As a part of our routine we generally devoted some portion of the evening to an examination of the fossils collected during the day. As I was checking one fossil after another on the day being described, I suddenly found something that looked suspiciously like a portion of a *Lystrosaurus* skull. Perhaps this story will sound contrived—to wish for *Lystrosaurus* in the afternoon and to see it in the evening. But that is exactly how it happened. The more I looked at the fossil, the more I became convinced that here, at last, was definite evidence of *Lystrosaurus*. The specimen wasn't much to look at, but it was distinctive, for

FIGURE 15. An historic fossil. A portion of the right side of a *Lystrosaurus* skull, with the tip of a tusk visible. Collected by J. A. Jensen on December 4, 1969. (See figure 10.)

it consisted of a portion of the right maxillary bone, the border showing the sharp downturned edge so characteristic of the front of the skull in this reptile, and within the bone, its point just protruding beyond the alveolus, or socket for the tooth, was very definitely the end of a tusk!

Jim Jensen had dug the specimen out of the sandstone cliff that afternoon, and associated with it he had found a leg bone—the ulna of the forelimb. Now that we had a portion of a *Lystrosaurus* skull at hand, with a limb bone, I could see that some of the other leg bones in our collection were probably those of *Lystrosaurus*.

We were all stimulated by this sudden insight into the nature of the fossils—very much so. All, that is, except one member of our larger scientific group—Isak Rust from South Africa. He took it with the greatest composure, for, as he pointed out, since the geology around us was so very similar to that of South Africa, why should we not find *Lystrosaurus*, one of the very common fossils in the African scene? Nonetheless, the rest of us were excited! Here, with these small bones of land-living reptiles, we had indisputable proof that in early Triassic time Antarctica was connected to Africa. This connection, so important to the concept of a former continent of Gondwanaland and for so many years a matter of conjecture, now came into the realm of solid fact.

LYSTROSAURUS IN AFRICA

Isak and his talk of Africa took me back seven years to memories of the Great Karroo—that immense shallow basin occupying thousands of square miles of desert, semidesert, and grassland to the north and east of the Cape of Good Hope, well on the way to Johannesburg. To paleontologists the Karroo is justly famous for its remarkable fossil

FIGURE 16. A skull and lower jaw of *Lystrosaurus*, in side view. The skull of *Lystrosaurus* is so distinctive that even a fragment, such as the fossil shown in figure 15, can be identified with certainty. Its position is shown by the diagonal lines.

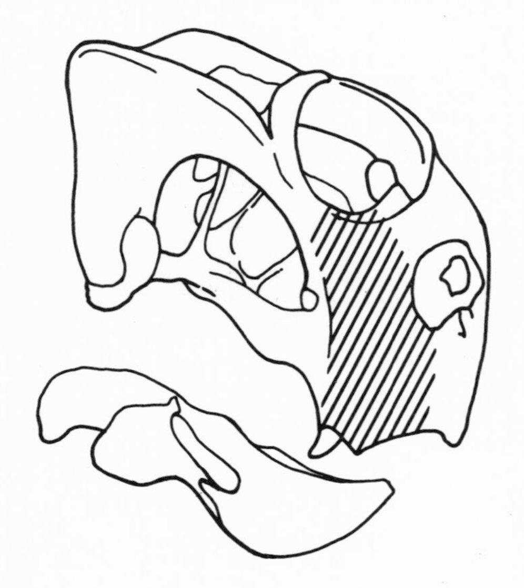

deposits, because in the concentric bands of sediments that circle the Karroo basin, like painted rings around the inner surface of a shallow bowl, there are immense numbers of fossil amphibians and reptiles, a petrified record giving silent testimony to the course of evolutionary history some 200 million years and more in the past. And that was why I had been in the Karroo, seven years in time and thousands of miles in space removed from Antarctica.

It was with a sense of pleasant anticipation that I set out with James Kitching, of Witwatersrand University, for a small exploratory trip across the length of the Karroo basin. To James it was an old story; he had had many years of experience in collecting Karroo fossils, but to me it was to be a journey into a new world, a world of fresh and exciting discoveries, a world of strange new sights and sensations. To me it was to be a rare and much-appreciated privilege, to enter this land that had yielded so many important fossils, a land across which numerous paleontologists had searched and dug in the years past.

We drove out of Johannesburg on a bright winter morning in June and headed toward the south and west. We were in a light truck, fitted out for field work, so we were ready to go anywhere that our inclinations and our search for fossils might take us; along the highways, out onto country roads, and across the fields. For the moment we were on the highway, traversing a landscape of singular charm and beauty. I recalled the opening lines from Alan Paton's powerful and emotional novel *Cry the Beloved Country:* "There is a lovely road that runs from Ixopo into the hills. These hills are grass-covered and rolling, and they are lovely beyond any singing of it."

Such was this land in which the winter fields lay brown and black beneath a benign sun, and across the fields moved groups of grazing cattle. Rather than the grass-covered hills beyond Ixopo, here sandstone cliffs towered above fields and valleys in processions of red and yellow and white walls. It was a land of many peoples and many troubles. It was a land to view with an eye adjusted to vistas of natural beauty seldom equaled upon this earth, while the mind pondered the problems of the varied peoples who inhabited these lovely valleys.

Toward the middle of the day we came to the "Golden Gate," where high cliffs of Upper Triassic sandstones—the Red Beds, so named because of their striking colors, and above them the white Cave Sandstone, so named because the cliffs in these rocks are frequently undercut and pierced by large rock shelters and caverns—make a natural portal to the land beyond. While we stopped to take some pictures—for one could not resist picture-taking in a country such as this—an African came walking up the steep road, pushing a bicycle, on the handlebars of which was tied the carcass of a big porcupine. To the man with the bicycle the porcupine was a prosaic enough object—something to take home and eat, I suppose—but to me it was more than prosaic, because this was my first encounter in the field with African wildlife, albeit the animal was very dead. In time I was to see many African mammals, of all sizes and very much alive, but I would never forget this first sight of the porcupine on the front of the bicycle, its enormous quills thrusting out in all directions.

The rocks in South Africa, from which so many fossil reptiles have been collected, have been divided into a series of successive zones, named after certain reptilian genera characteristic of the several zones. From the bottom up there are the *Tapinocephalus, Cisticephalus,* and *Daptocephalus* zones (following the revised nomenclature published by James Kitching), all of late Permian age, and above them, the *Lystrosaurus* and *Cynognathus* zones, these being of early Triassic age. We were journeying into the *Lystrosaurus* zone, near the town of Harrismith.

The early morning sun the next day cast long shadows across the frosty grass, and one huddled in a sweater and felt cold; in retrospect, that early winter morning in South Africa seemed colder than the mornings in Antarctica. The smoke from the morning fires in the African houses rose straight into the still air for a few feet and then spread horizontally, to make a low, gray layer against the colorful cliffs. The delicious smell of the smoke permeated the air. Here and there, across the fields, were large holes, where aardvarks had been digging during the night.

Although it was cold, the air rapidly was getting warmer. There was the promise of a fine winter day, when the sun feels good during the midday hours and when it is a joy to be out of doors. Far and wide, among the trees across the fields, could be heard the constant calling of little doves. The land was big and stretched for miles in almost every direction; steep escarpments jutted up to form far battlemented horizons. It was a scene that demanded attention.

Yet only a certain amount of attention could be given to the distant view. We were looking for fossils, and James already was walking across some broken gullies, or "dongas," scanning the banks with an experienced eye. Of all the fossil hunters with whom I have been associated, none is the equal of James Kitching. He has an eye for fossils

FIGURE 17. James Kitching, the doyen of South African fossil collectors, searching for *Lystrosaurus* skeletons near Harrismith, Orange Free State. He found several on that crisp winter morning.

that is truly phenomenal, and his ability to discover fossils in the rocks is justly celebrated on more than one continent. One story (and I do not doubt it) tells of James on a train, seeing a fossil skeleton in a railroad embankment as the train sped through a deep cut. James got out at the next station, hired a car, returned to the spot, and collected the fossil. Certainly, during those days that I was with him in the field, he found about eight or ten fossils for every one that I saw.

So with James scouting the land for fossil reptiles, I was not inclined to linger too long in admiration of the distant scene. Although this was no contest, at least I hoped that my showing would be not altogether feeble when compared with the riches that I knew James would turn up from the earth. Therefore I joined him in the hunt.

It was not long until we found our fossils. James found a nice skeleton, curled up with the head pointed toward the tail. Soon after I found one—not so nice as the first specimen but nonetheless a part of a skeleton with the skull, and something to give me a thrill. It was my

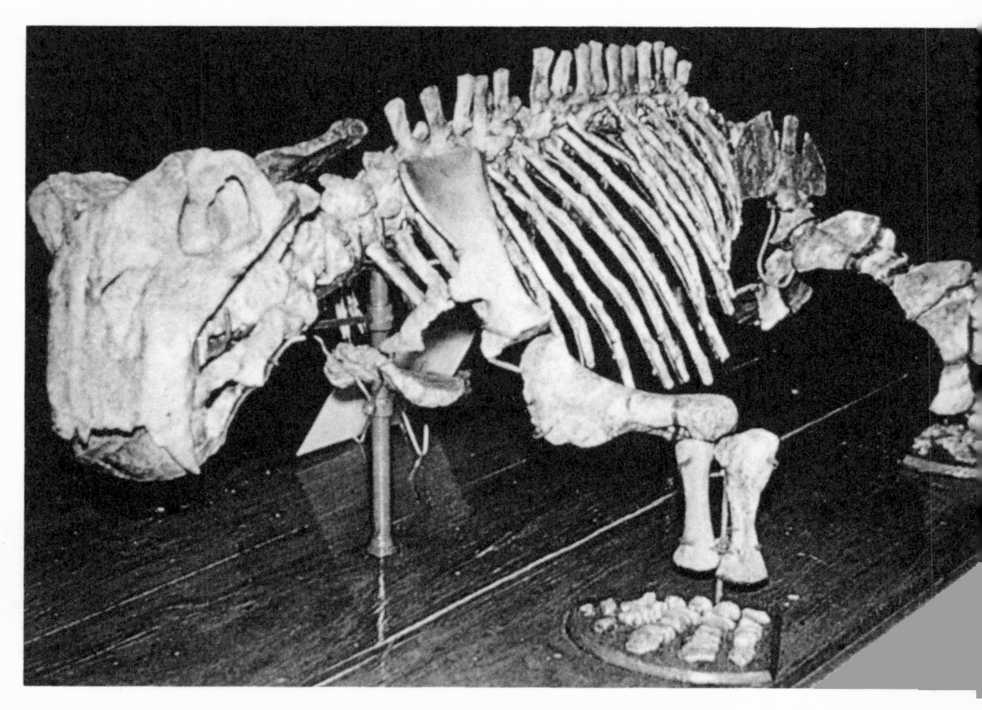

FIGURE 18. A skeleton of *Lystrosaurus* from South Africa. Here is the bony framework over which the reconstruction (see figure 12) was made.

first fossil specimen in Africa, and it was *Lystrosaurus*. One could not mistake it, with the downturned snout, apparent even though the skull was largely covered by rock, with the tip of a tusk showing, and with the short limbs and the widely expanded ribs of the stout body closely pressed to the back of the skull.

Through the morning we carried on our search, and we continued to find more fossils, some very fragmentary and some in rather nice condition. And almost all of them were *Lystrosaurus*. These fossils were ubiquitous. One got the impression of *Lystrosaurus* living in Triassic Africa in huge concentrations as herbivores so often have lived. However that may have been, this reptile obviously was very abundant during those days some 200 million years ago, when reptiles ruled the earth. *Lystrosaurus* was indeed a very successful reptile.

I need say no more about our African fossil hunt. We found more *Lystrosaurus* and other reptiles besides, and in the other fossiliferous zones of the Karroo, above and below the *Lystrosaurus* level, we found other fossil treasures. But it was *Lystrosaurus,* to single out one genus among Karroo fossils, that occurred in the greatest abundance.

This was made especially evident to me when I looked at the collections in the Bernard Price Institute for Paleontological Research at Witwatersrand University. There one will see case after case and in each case drawer after drawer filled with *Lystrosaurus* specimens. The Lower Triassic outcroppings in the Karroo have been a great hunting ground for the seeker of *Lystrosaurus,* and they will continue to be for countless years to come, as the inexorable processes of erosion wear away the rock and the soil.

LYSTROSAURUS IN INDIA

But memories, as I examined the Antarctic *Lystrosaurus,* were not restricted to Africa. I was carried back in my mind some five years, to think of a train crossing the flood plain of the Ganges, going from Calcutta to Asansol. Although it was January, the day was hot, and we felt it in our compartment, which did not benefit from the comforts of air conditioning. The Indian gentleman opposite us removed his shoes and socks, and made himself comfortable in his own way, by pulling his legs in beneath him on the seat. It was satisfying, no doubt, but to those of us who had grown up in a culture of chairs and tables, it appeared to be far from relaxing. Nevertheless, he maintained this position for mile after mile and devoted himself to an obviously engrossing book. Outside we could watch the bullock carts and trucks and an occasional passenger car, journeying along the road that paralleled the tracks. The sun shone intensely, and the shadows were black beneath the trees.

We were three, in addition to the Indian gentleman, together in the compartment: my wife and myself, and Dr. Pamela Robinson, of the University of London. And it was thanks to Pamela that we were there, embarking upon an adventure that was new to us. Pamela, who has worked extensively in India under the auspices of the Indian Statistical Institute in Calcutta, had arranged with Professor Mahalanobis, the famed statistician, and then director of the Institute, for us to come to India to see some of the Triassic of that fabled land. So there we were, riding in a hot train to the northwest, to spend a little time in the Panchet Formation of the Raniganj coal field, where fossils are to be found in those Lower Triassic beds.

The station at Asansol was typical of railway stations in India. As our train pulled in, we could see that the platforms were crowded with masses of brightly garbed people—men in white pajamas and punjabis, turbaned men in scarlet jackets and white dhotis, diaperlike garments so commonly worn by many Indians, men in dark business suits, youths in slacks and sport shirts, bearded Sikhs, women in saris and women in the trousers worn by the Punjabi ladies of the north. All of these people were waiting patiently on the platforms, some of necessity standing or sitting in the hot sun, others enjoying the shade, where they could find it. And as the train ground to a stop, we left our hot and dusty compartment to join the multitude on the platform.

Soon we were inside the station, where we encountered some striking and, to us remarkable, works of art. There were two statues of Sarasvati, the goddess of learning, prepared for the festival that was to do her honor. They were exquisitely made and painted, and they depicted Sarasvati clad in flowing robes and dancing, it would seem, on a large-petaled flower, in one of the graceful steps so characteristic of Indian dances. These images, we learned, were made of clay, over a straw manikin, and were painted and glazed, to give them a certain degree of permanence. They excited our greatest admiration; they were highly sophisticated and far from amateurish. One could not help but be impressed by the work and the talent that had gone into such beautiful but comparatively transient objects.

We spent some time in admiring these appealing statues, but at last we bid them adieu and went outside to the street. There, in front of the station, a red brick castle of distinct Victorian aspect, were two of our Indian colleagues waiting for us with a couple of jeeps—Bimelendu Roychaudry of the Statistical Institute, who had spent some years in the United States, where he obtained his doctorate in geology at the California Institute of Technology, and Mr. P. P. Satsangi, of the Geological Survey of India. As we were being greeted I could not help glancing out of the corner of my eye at a large, hirsute man coming along the street, his head covered by a turban, his white punjabi shirt partially hidden by a dark vest, and behind him a woman in a saffron-colored sari, carrying a baby on her back, while to one side was a lean man between the shafts of a rickshaw, waiting for customers. India was still new and strange, and the local color of the passing scene frequently distracted one from the business at hand.

We boarded our jeeps and departed through the streets of Asansol for our field camp. Past people carrying bundles on their heads, past laden bullock carts, past noisy trucks we went, threading our way through crowded streets and roads, until we reached the open country, the roads of which were still filled with the varied people and vehicles of India.

FIGURE 19. Camp in the Damodar Valley of India, about 150 miles northwest of Calcutta. Susunga Hill in the background.

Our camp, outside the little village of Tiluri, south of the Damodar River, was an interesting affair. We had kindly been given permission by the authorities to occupy an unfinished complex of small buildings that were to be a health center and school. So we set up our cots in the bare rooms of what were to be the nurses' quarters and used the largest building of the group for a mess hall and gathering place. There were perhaps ten or a dozen of us, since in India all projects, even fossil hunts, require for the principals involved a satellite ring of servants and helpers. Thus, in addition to Pamela, Bimelendu, Satsangi, my wife, and myself, there was Debraj, the cook (and a very good cook he was), and his helper, a driver for each of the jeeps, and a general factotum named Ram, a huge man with a fierce moustache.

Every morning, as the red sun came up over Susunga Hill beyond our camp, Ram would come in with tea. Every day, as we explored the sediments of the Panchet Formation, Ram was there with us, carrying extra geological hammers and other supplies in a haversack on his back. Every noon, when it was lunchtime in the field, Ram would appear with a contraption of nested tin bowls containing curries, and thermos jugs of hot tea. (Hot curries and hot tea sometimes seemed a bit torrid, after a sweltering morning under the Indian sun.) Throughout the day Ram did what he could to make life pleasant for all of us; he was a gentle and helpful soul and his cheerful character quite belied his formidable physique and his terrifying moustache.

The Panchet beds consist in part of brown sandstones that become almost black when weathered and are frequently in the form of low cliffs, capped by large, weathered sandstone blocks. Beneath the sandstones, or associated with them, are reddish and brownish shales and clays, and it is within these comparatively soft sediments that fossils are found. The sandstones and clays of the Panchet beds, which form in part large jungle-covered hills that dominate the landscape, are also exposed in small, eroded stream valleys, known as "nullas" (the equivalent of the dongas in Africa), and it was to the nullas that we would make our way each day from our camp near Tiluri. The nullas are, of course, erosional interruptions in a countryside where every possible square inch is devoted to fields and to villages. Therefore, we saw much of village and country life in our passages from camp to the collecting grounds.

From a little distance the domed towers of a temple, looming through the morning mist, showed us the location of Tiluri. It was a picturesque and impressive temple, and distance lent enchantment to its form and bulk. Near at hand its details detracted somewhat from the mystery of Asia; there was a certain scaling of the walls, even the sight of brick where the plaster had fallen away, and a certain fading of the paint that lent a rather scruffy appearance to the structure. Nonetheless it was a most interesting building, and we enjoyed looking at it every time we went through town. The streets of Tiluri, of packed earth, were narrow and crooked, and bordered at their very margins by the houses of the town. The houses, of earth, had painted or whitewashed walls, often much faded and more often than not decorated by numerous pancake-sized patties of cow dung, slapped against the walls to dry in the sun. These were fuel for the villagers. The roofs of the houses were various, of thatch, of red tiles, or quite frequently of corrugated iron.

There was always life in the narrow streets, human and animal. People were going to and fro, many of them carrying bundles. Women and girls often were burdened with young children, carried on their backs or on their hips. Men would be walking along or sometimes clustered in little groups, standing or squatting and talking in an animated fashion. Cattle wandered freely in the streets, as did yellow dogs. Here and there a goat might be tied up by a house. Through this melange of busy people and animals our jeeps would honk their way, while we sat and prayed that we would not hit any living thing.

Beyond the village we were in the countryside again and finally at the nullas, where we were to study the geology and look for fossils. In some of the nullas there would be a little water, where frequently country women would be washing clothes. And in one nulla a group of young men were wading through the shallow water and probing with their hands for fish. At one place we found an outcropping of weathered sandstone with a design of shallow grooves cut into it. This, we were told, was a game called "tiger," played by the country children, in which the grooves form a sort of base or "board" on which pebbles are moved about—something on the order of our various checker games.

The most exciting thing we saw on our first day in the field was a fight between two water buffalo. These useful animals ordinarily are very placid, and one can walk among them with no concern. So we were treated to an unusual and a rather terrifying sight.

We had just climbed out of a nulla, attracted by the noisy shouts of many people. There were the two animals, in violent combat, with the villagers swarming around and trying to stop the fight. (After all, the two buffalo were valuable items in the economy of the country people, and nobody wished to see either of the animals injured.) But the efforts of the peacemakers were of no avail. Each of the two buffalo was truly intent on killing the other one. They fought with single-minded intensity and with the brute force of a thousand pounds or more on each side. Repeatedly one buffalo knocked the other one down and tried to gore him, all in spite of the sticks and yells of the surrounding crowd. Finally, one buffalo that was down struggled to his feet and made off across the fields, with his adversary in hot pursuit. That is the last we saw of them; we never learned the outcome of the affair.

Our work in the nullas involved a good deal of tramping, a lot of observation of rocks, with concomitant discussion among us as to the meaning of many things we saw, and of course a constant search for the evidence of fossils. One morning, shortly after we had begun our campaign, I was lucky. In a bank right in the middle of a nulla I found a *Lystrosaurus*.

It was, in effect, in a little mound, an erosional remnant that had been left isolated in the nulla. I walked around this small feature, and there on one side I saw a skull partially exposed. Of course I called my companions, and we all gathered around to view the specimen and to

FIGURE 20. Excavating a *Lystrosaurus* skeleton in the Damodar Valley, with the usual audience. Facing the camera: Dr. Pamela Robinson of the University of London. Standing with backs to camera: Dr. Bimelendu Roychaudhury of the Indian Statistical Institute and Mr. P. P. Satsangi of the Geological Survey of India.

examine it. This did not take long. Some scratching with awls, especially adapted for the purpose, soon revealed a beautiful specimen of *Lystrosaurus*, the skull and jaw complete, and behind the skull, the skeleton, curled up with the tail near the head—a pose in which one frequently finds such fossil reptiles preserved. We saw that we had our work cut out for us for the next day or so.

When we got back to our camp we made our plans and assembled the equipment and materials that we needed—picks and shovels, digging tools, brushes, shellac, plaster of Paris, and coarse cloth. On the next day we went to work. We dug around the specimen, exposing it just enough to indicate the limits of the skeleton. Any bone that was exposed we treated with thin white shellac. This hardened the bone. We also soaked the matrix around the bone with shellac. When the shellac was dry, we covered the specimen with thin tissue paper, made a creamy mixture of the plaster, dipped strips of cloth in this mixture and plastered the cloth over the specimen. When these "bandages" hardened they made a tough coating, like a cast that physicians place over a broken limb, and thus prevented the fossil skeleton from disintegrating. We plastered some sticks onto the skeleton that was so covered, to make reinforcing splints. Then we undercut the specimen, turned it over, and applied the same methods to what had been the underside of the skeleton. The fossil was then ready to be moved.

During all of this procedure, we had an interested and frequently large audience of country people—young boys, youths, and men, many of them enveloped in togalike garments, who sat around us in a semicircle and watched us work. It was something very new and unusual in their lives, so naturally they followed our every move with great interest.

Finally we were ready to take the fossil to our waiting jeep. It had now become rather heavy, encased as it was in its protective plaster cast. We rolled it onto a large piece of burlap, brought the edges of the

FIGURE 21. The Indian *Lystrosaurus* skeleton, properly encased in its protective jacket of plaster, being carried to the field vehicle. Ram, our burly factotum on the front end of the pole, does the work of three, while persuading some water buffalo to make way for the fossil.

burlap together on top of the plaster cast, tied them together, and through this improvised sling thrust a long, stout pole. Then a half-dozen of us, putting our shoulders to the pole, lifted the fossil clear of the ground, and struggled across the nulla and up its side, to the jeep. On our erratic, grunting, and sweating progress we had almost literally to kick an unemployed water buffalo out of our path. He was very docile and made way for us with no protests offered.

BACK TO ANTARCTICA

Thus I had been fortunate enough to collect *Lystrosaurus* on three continents: in Africa amid varied people and animals living in broad valleys between colorful sandstone cliffs, in India amid fields and villages and their inhabitants that filled the plains between the hills, and in bleak Antarctica, where we who searched were the only living things. Why Antarctica?

We certainly were not there for the express purpose of finding *Lystrosaurus,* although the possibility of discovering this interesting reptile was in our minds when we went. We were there because it seemed that we might have some chance of finding Triassic amphibians and reptiles, and this was a chance worth the taking.

For some years I had hoped that someone would go to Antarctica to look for fossil tetrapods—the four-footed animals that in late Paleozoic and early Mesozoic times were represented by amphibians and reptiles. In fact, a few years before, it had looked as if James Kitching might be the man to carry out this project, and no better man for the task could have been found. As mentioned before, James had an eye for fossils in the field that is almost beyond belief, so that if anyone were to find early tetrapods in the wastes of the Antarctic, James was the man. But circumstances prevented his going, and then for some years the hope for such a quest was dormant.

As briefly mentioned in the Prologue, in the austral summer of 1967–1968, Peter Barrett, a New Zealander at the time connected with the Institute of Polar Studies, was making a geological reconnaissance of the Transantarctic mountains in the general vicinity of the Beardmore Glacier. On a day in December of 1967, he, with a fellow scientist, was exploring the sandstones of the Fremouw Formation, high on the side of Graphite Peak, about 30 miles to the east of the great glacier. He certainly was not looking for the remains of ancient tetrapods; his training and his purpose were in the field of geology, and he was applying himself to his purpose by ardently examining and studying the rocks that were exposed in this region. Something caught his eye—a strange-looking shape enclosed within the sandstone of the Fremouw beds, so he and his companion gave it their attention. With their hammers and picks they were able to break the specimen out of the rock, and when it was free, it seemed to them that perhaps this was a bit of fossil bone. Could they be sure?

When they got back to the four-man camp that was the base for their operations, they showed the specimen to another member of the group, Ralph Baillie, who had the benefit of some training in vertebrate paleontology. Ralph immediately saw that the specimen was a fossil bone, so he treated it with such materials as he had at hand to preserve it for future study. Naturally all four men were intrigued, one might say excited by the specimen. They realized that here was something unprecedented in the annals of Antarctic exploration.

In January 1968 the storms began to increase in intensity, as they do in Antarctica at that time of year, and the four men returned from their field camp to McMurdo base and from there to New Zealand. Ralph was custodian of the fossil, which he carried carefully nested in cotton in a small tin box. When they landed at the airport in Christchurch, where it was necessary to make a customs declaration, Ralph wrote down that he had a fossil bone in his luggage. If he had only said "rock," everything would have been all right, but that word "bone" excited the customs officials. New Zealand quite justifiably has a strict embargo on such organic materials, but a petrified bone some 200 million years in age is about as inert an object as may be imagined. Nevertheless the customs declaration had the word "bone" on it, so the fossil was confiscated, sequestered, sealed, and placed in a very cold deep freeze, there to remain in dark, frigid seclusion until the Antarctic explorers left New Zealand to return to the United States. Then the innocent fossil was returned to them—and good riddance in the eyes of the customs officials.

As has been mentioned, the specimen came to me for identification, and I saw that it was a piece of jaw of a labyrinthodont amphibian, the labyrinthodonts being ancient amphibians that arose from air-breathing fish ancestors. The first labyrinthodonts appeared at the end of the Devonian period, about 340 million years ago, but theirs was a long-lived group, and so they persisted through the Triassic period, the geological division with which we are presently concerned, to become extinct at its termination, a little less than 200 million years ago. A closer identification than this was not possible; the specimen was too incomplete to give a clue as to its generic identity. Nonetheless, this was sufficient to indicate a discovery of major importance, so the fossil gained a great deal of attention, not only in scientific circles, but also in the public press.

The National Science Foundation was more than pleased by this discovery. Consequently I made a point of stressing the importance of searching for more fossils. Where there was one, it seemed logical to suppose that there should be more. The logic was convincing, and as a result the foundation decided to sponsor a search for Triassic vertebrates in the Transantarctic mountains, even though such a fossil hunt would be very expensive and might wind up with negative results. Still, it seemed like an opportunity of which all advantage should be taken.

MODERN ANTARCTIC FOSSIL HUNTERS

A fossil hunt in the Transantarctic mountains would require massive logistics: transportation of personnel, equipment, and supplies from McMurdo Station to the field by a giant C-130 Hercules plane, close helicopter support during the field season. Why not make it more inclusive, to take advantage of the facilities that would be necessary for the work? The decision was made, and the 1969–1970 program for work in the Transantarctic mountains grew into a large and comprehensive effort, embracing many phases of geology and paleontology. That is how the Beardmore Camp, as it was called, grew into the large operation that already has been briefly described. And that is how I found myself in the Antarctic.

Our group of about 20 scientists was under the general supervision of Dr. David H. Elliot, a veteran Antarctic geologist connected with the Institute of Polar Studies. It might be mentioned that David is an Englishman. It also might be mentioned that David is an excellent leader and was affectionately admired by the people who worked with him.

Our group left the United States in late October and flew by military aircraft to Hawaii, and on to Christchurch, New Zealand. There some of us were held up for a week or so by bad weather in the Antarctic. We made six attempts to fly from Christchurch to McMurdo Station, and on the sixth venture we made it. Then we were held up at McMurdo Station for three more weeks by bad weather—this being the worst spring, weatherwise, in Antarctica since Scott made his final dash to the Pole in 1912.

FIGURE 22. Logistics in Antarctica. A Hercules C-130 with supplies for our camp. The Hercules, with its load-carrying capacity and its ability to land and take off on unprepared ice fields, has revolutionized Antarctic exploration and field work.

Finally, on the twenty-second of November, our party boarded a Hercules and made the flight to our camp, which had been set up by the Navy Seabees. We had a steady journey in brilliant sunshine along the Transantarctic mountains, and across the immensity of the Beardmore Glacier, which from the air is more than impressive—it is terrifying. We landed at our Beardmore Camp that afternoon.

In spite of the bright sunshine, there was a heavy wind blowing when we alighted from the plane, a wind that carried clouds of snow across the vast ice field that was to be our immediate environment. We were glad to get into the Jamesway hut where we were to live. Even inside the Jamesway it was uncomfortably cool, so strong was the wind. Therefore we set to and spent the remainder of the afternoon cutting snow blocks and piling them against the sides of the hut to afford additional protection against the wind and the cold.

The camp consisted of four Jamesway huts, one for the scientific personnel of the camp (commonly called Usarps, an acronym for United States Antarctic Research Program); one for the Navy men—

pilots, mechanics, and support personnel; one a cook shack and mess hall; and one a radio shack. A generator had been set up, so we were supplied with electric lights. Indeed, it was a comfortable, almost a luxurious camp, to be established in the interior of Antarctica, within 400 miles of the Pole.

Something should be said about the Jamesway huts. These are a portable modification of the Quonset hut, which became famous during World War II. Instead of a steel frame, the Jamesway has a wooden one, nicely designed so as to be folded and disassembled, and packed into a relatively compact space for shipping. Instead of a metal covering, the Jamesway has a heavy, insulated covering of fabric. The floor is of heavy plywood, strongly braced beneath. These huts were flown from McMurdo Station to our Beardmore campsite in a Hercules, and within the course of a few hours each was erected and made ready for occupancy.

Each hut was heated by a Preway oil heater and on the whole was very comfortable. It was interesting, however, to note the variability of temperatures within the hut. At floor level, water if spilled would freeze; at waist level the temperatures were usually about 40 degrees; at shoulder level a comfortable 70 degrees was common; and at the top of the arched hut the temperature was commonly 90 degrees and more.

Our first afternoon in camp was, as mentioned, a bit uncomfortable, and much of our effort was expended in insulating the sides of the hut with snowblocks. On the next morning, which was a Sunday, almost all of the camp elected to stay at home in order to get properly settled. There was much to be done in the way of arranging gear and accommodations. David Elliot, however, decided to go over to the nearest rocks for a little preliminary exploration. One or two other members of our group accompanied him.

The "nearest rocks" were in the long ridge known as Coalsack Bluff, with which the reader already has been made familiar. This brings up a question: Why was our camp within a few miles of Coalsack Bluff? The answer is that our presence at that particular place was more or less fortuitous. It was more by accident than by design that we were near Coalsack Bluff, which was destined to become the scene of our operations through the Antarctic summer. All of which illustrates the role of chance and the factor of serendipity in the scientific solutions of problems, especially when one is doing geological or paleontological work in the field.

Our camp had been sited there by David principally because it was a good place for the big Hercules planes to land. He might have placed it 30 miles away at Prebble Glacier, but he didn't. He had flown over this region, and it seemed to him that Walcott Névé, the great ice sheet on which our camp was located, was perhaps as good a place as any to serve as the base for our operations. The ice was extensive, it was reasonably smooth, and it was free from crevasses. Some trial landings with a Hercules showed that it was a feasible spot for landing the big ski-equipped planes, which pleased the admiral in charge of Navy operations, because during a previous field season one of the planes had seriously damaged a ski in attempting to land in this part of Antarctica. (The cost of a ski for a Hercules, it might be added, runs into six figures.) And so we were camped on the ice, with the magnificent heights of the Transantarctic range some 20 miles or more away to the south. Since we were to have helicopters for close field support such distances had little significance; in a helicopter one could go from camp to a field locality in the mountains within a few minutes or perhaps half an hour.

Coalsack Bluff, curiously enough, had not figured at all in our preliminary plans for a fossil search. We had pinpointed a score or more of possible fossil localities in the Transantarctic mountains, where good exposures of the Fremouw Formation were evident, and it was to these localities that we intended to devote our attention, with the helicopters to take us back and forth.

So on that Sunday morning, before we had had a chance to get our field program organized (the helicopters had not as yet arrived), David Elliot decided to go over to Coalsack Bluff, mainly for want of something better to do. This is where serendipity enters in. He and his companions took off with a motor sled pulling a light Nansen sled and we waved them a goodbye.

At Coalsack Bluff the element of chance entered in. It happened that David was having an argument with one of his associates, Jim Schopf, a paleobotanist, concerning the age of the sandstones that were exposed in the side of Coalsack Bluff. The lower portion of the bluff, already described, is composed of Permian shales and coal seams (the coal seams having inspired a New Zealand sledging party that came through there some years before to apply the name "Coalsack" to this ridge), and the question was how far up such Permian rocks extended. David climbed to the sandstones high up on the Bluff, because he thought they looked as if they should belong to the Triassic Fremouw Formation. When he got there he satisfied himself that they did, and while poking about on the cliff faces he found two tiny fragments in the rock of what seemed to him to be fossil bone.

So it was that at lunchtime the party came back, the motor sled puttering away ever louder through the clear air as the distant black speck on the shining ice field resolved itself into sleds with people on them. David came into the Jamesway with an air of suppressed excitement, to exhibit the two tiny fragments that he carried in his hand. And I could confirm their nature as fossil bone. Consequently after lunch a considerable group of us set out for the bluff, the motor sled again chugging across the ice, pulling a retinue of expectant people.

We spread out along the little cliffs when we finally reached our destination, and within minutes we knew that we had a fossil bonanza. All along the cliffs we found fossil bones sticking out of the cross-bedded sandstones. The more we looked, the more we saw. It was evident that here we would be working—at least for quite a time into the future of the field season. Our plans for distant trips to the Transantarctics in helicopters were scrapped for the time being. We shook hands around and went back to camp in a very happy frame of mind.

This brings us back to the beginning of the story, and explains how it was that we were on Coalsack Bluff, searching for and collecting fossil bones, and how it was that I was wishing with all my heart we would find a *Lystrosaurus*—which we did.

On the very evening that we definitely recognized *Lystrosaurus* among our fossils—as represented by the piece of skull collected by Jim Jensen—a Hercules landed bringing Dr. Laurence Gould, the dean of Antarctic geologists, for a visit to our camp. Larry Gould viewed the *Lystrosaurus* with very evident excitement; he immediately appreciated its significance. And when he returned to McMurdo Station he radiotelephoned the information to the National Science Foundation, so that within a few days the news had flashed across the world, to be announced in the newspapers and news weeklies of many nations. Thus *Lystrosaurus* became known to countless numbers of people, including nonpaleontological scientists, who had never heard of it before.

Almost anyone, no matter what his background might be, could appreciate the meaning of *Lystrosaurus* in Antarctica, once a few facts were made clear. Here was a land-living animal of ancient age, virtually identical to animals in Africa and Asia. Somehow *Lystrosaurus* and the tetrapods associated with it in Antarctica had made their way to that continent from Africa on dry land. Here, it seemed, was proof positive of the former connection of Antarctica with Africa, and probably other southern-hemisphere continents as well; here, it seemed, was one of the most telling facts ever discovered to indicate the probable reality of an ancient Gondwanaland, and of subsequent continental drift. Here was a discovery of undoubted significance concerning life on the continents of long ago, and the bearing of the distribution of that life on former continental relationships.

The discovery of *Lystrosaurus* and associated land-living vertebrates at Coalsack Bluff revealed new aspects in the interpretation of earth history. It was a discovery of the utmost significance, since it opened a window, so to speak, giving us a glimpse of animals that had, many millions of years ago, inhabited a continent almost as large as North America. And by showing us something of this ancient life, clues were revealed as to the relationships of Triassic Antarctica and the animals that had lived there when it was a land of warmth, to the various other continents and their inhabitants of 200 million years in the past. The work at Coalsack Bluff afforded a most significant and exciting glimpse of long-distant years, but it was only a beginning. How would future work among the snow and the glaciers of the Transantarctic mountains augment the discoveries that had been made in this one short field season?

McGREGOR GLACIER

As an answer to this question a second foray into the field was planned for the following year, the 1970–1971 season. This was to continue work originally planned for the first year of fossil hunting, but which had not been carried out owing in large part to the inclement Antarctic summer of 1969–1970, a summer of storms that had greatly reduced the number of days on which we could venture out from camp. It had been planned, after working at Coalsack Bluff, to devote the latter part of the 1969–1970 field season to collecting in the vicinity of McGregor Glacier, across the great Beardmore Glacier and some 150 miles to the east of Coalsack Bluff. Alas! The best-laid plans of the fossil hunter often go astray, for unforeseen reasons. So we never did get to McGregor Glacier during that first season.

And that's how it was that in the autumn of 1970 the Hercules aerial workhorses carried another crew to the Transantarctic mountains, to continue the search—this time at McGregor Glacier. The principal paleontologist was James Kitching; at long last his remarkable talents in the field, so amply displayed in the Permian and Triassic Karroo beds of

South Africa, would be applied to the cliffs and outcrops thrusting through the ice of Antarctica. With him were John Ruben, of the University of California, and for a short time, Thomas Rich, of the American Museum of Natural History in New York. As during the previous year, there were also geologists other than the vertebrate paleontologists participating in the work at McGregor Glacier.

History repeated itself. The party settled itself in camp at McGregor Glacier, and on the first day of field work a fossil was discovered—again by a geologist rather than by one of the paleontologists. James Collinson, who had been a member of the previous year's group, devoting himself particularly to studies of the sequence and relationships of rock strata in the Transantarctic mountains, stumbled onto a most intriguing fossil in the Lower Triassic sediments of the Fremouw Formation—only a few miles from camp. It was an enigmatic-looking specimen, an impression in the rock, and it required the keen and knowing eye of James Kitching to identify it as the impression of a skeleton of *Thrinaxodon,* a small carnivorous mammallike reptile, perhaps comparable to a weasel in size and probably, like its modern mammalian counterpart, an aggressive little predator. The skull in *Thrinaxodon* is that of an active carnivore, its sharp teeth are well adapted for grasping and biting and tearing, and its strong legs, drawn well in beneath the compact body, indicate an animal capable of running across the land with considerable agility. *Thrinaxodon* is quite characteristic of the *Lystrosaurus* zone in South Africa.

As had happened a year earlier, excitement ran high, and within a short time the news of *Thrinaxodon* in Antarctica was appearing in papers throughout the world.

This discovery was but the first of many; during all of the weeks that the party was at McGregor Glacier many fossils were found. They came to light at several localities—at Thrinaxodon Col, obviously named because of the concentration of this interesting reptile; at Kitching Ridge, so named because here James Kitching found one thing after another, day after day; and at other places as well. Moreover, the fossils of 1970–1971 were of particular importance because they frequently occurred in the form of articulated skeletons—a contrast to the isolated bones that had been the rule at Coalsack Bluff. They were abundant, and they kept James Kitching very busy indeed.

FIGURE 23. A skeleton of the Triassic mammal-like reptile *Thrinaxodon* at McGregor Glacier, Antarctica. This, the first *articulated* skeleton to be found in Antarctica, was an exciting discovery. *Thrinaxodon* commonly occurs with *Lystrosaurus* in Africa; here was evidence for the same association in Antarctica.

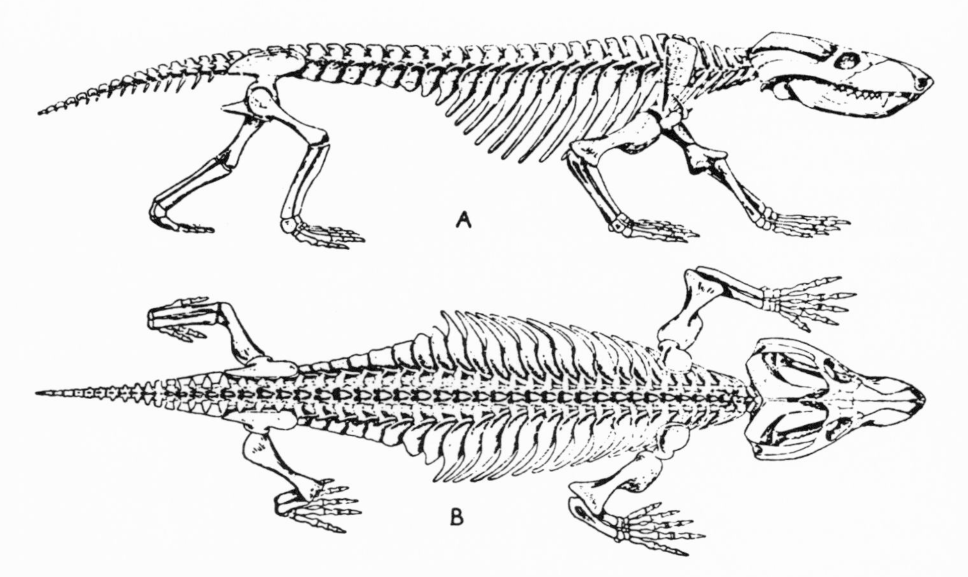

FIGURE 24. The skeleton of *Thrinaxodon.* This mammallike reptile is distinctive by reason (among other characters) of its expanded ribs. Thus *Thrinaxodon* in Antarctica could be equated with *Thrinaxodon* in Africa because of the peculiar rib structure, just as *Lystrosaurus* could be equated in the two continents on the basis of its distinctive skull. A. Side view. B. Top view.

The fossils from McGregor Glacier truly did supplement and augment the fossils from Coalsack Bluff beyond all expectations. More *Lystrosaurus* were discovered, quite identical with the *Lystrosaurus* of Coalsack Bluff, and in addition, of course, *Thrinaxodon.* Also there were found in the rock outcrops fringing the glacier the bones of a small, primitive reptile, *Procolophon,* of small protolizards, and of small amphibians, all very similar to animals associated with *Lystrosaurus* in Africa.

So if the discoveries of the first field season pointed to the close relationship between the Lower Triassic land-living animals of Antarctica and Africa, those of the second season truly clinched the evidence, most convincingly. There could be little doubt of it. During those years of early Triassic history, before the dinosaurs had come into being, various amphibians and reptiles of the same kind had lived in Africa and Antarctica. Their essential identity in the two regions, now so distantly separated and now having such antithetical climates, indicates almost beyond argument that these land-living vertebrates constituted a widely distributed association of animals, a fauna, living across 2,000 miles of what was then continuous land, formed by the conjunction of what is now southern Africa and what is now Antarctica. Throughout this region these ancient amphibians and reptiles evidently wandered without hindrance, just as quite recently, lions, leopards, and numerous antelopes spread across vast regions of the African continent, or as quite recently, cougars, wolves, deer, and bison likewise spread over much of the extent of North America. (*Lystrosaurus* with associated reptiles and amphibians also was present in India, as we have seen, and in China as well.)

FIGURE 25. James Kitching at McGregor Glacier, after having collected the *Thrinaxodon* skeleton. He has reason to smile.

This brings us to a very large subject, already briefly introduced, namely the relationships of past continents, and of the animals that lived on those continents. It brings us again to Gondwanaland, Laurasia, and Pangaea, for an examination in some detail of these ancient land masses.

II. GONDWANALAND, LAURASIA, AND PANGAEA

LYSTROSAURUS AND CONTINENTAL CONNECTIONS

Fossil remains of *Lystrosaurus* in Antarctica, in southern Africa, in India, as well as in western China are in a sense the paleontological beads from a broken string, each occurrence separated from the others by thousands of miles of deserts and grasslands, mountains, jungle, or open ocean. Yet the petrified bones of this ancient and interesting reptile are so nearly identical, wherever they occur, that it seems evident the several populations of *Lystrosaurus* were in early Triassic time inhabiting a region within which they could move freely, thereby maintaining an identity of form and function among themselves. There seemingly were no significant barriers to prevent the flow of *Lystrosaurus* genes back and forth; there were no mountain ranges or oceanic straits to bring about the development of isolated populations exhibiting the differences that one sees between separated groups of animals. For *Lystrosaurus* the early Triassic world in which it lived was one world, perhaps a world in which the limits of lystrosaurian distributions were on the whole measured by relatively modest distances, and perhaps only occasionally by sweeping arcs extending across many thousands of miles of latitude and longitude.

Of course, when one looks at a map of the modern world it is possible to suppose that *Lystrosaurus* theoretically might have extended its range from South Africa up through the African continent, across Asia Minor, and on into India and western China. Such an extended range of distribution is not particularly unusual in the modern world. For example, the leopard—a single feline species—not so long ago ranged from the Cape of Good Hope through Africa, across the Middle East and Asia, and well into Siberia. But how is one to account for *Lystrosaurus* (and associated land-living vertebrates) in Antarctica?

This reptile, although quite probably a lover of lakes and rivers, an animal that might be regarded as a sort of ecological miniature reptilian hippopotamus, was nevertheless a denizen of continental regions. It could no more cross great stretches of ocean than can a hippopotamus; it had to move from one region to another by a land route, albeit a well-watered path of migration. And what was true for *Lystrosaurus* was equally true for the small carnivorous mammallike reptile *Thrinaxodon,* living its small-scale predatory life in the undergrowth of early Triassic landscapes, and for the small primitive reptile *Procolophon,* a sort of analogue of the modern horned lizard, as well as for various other ecological companions.

A few decades ago it was common, as we have seen, for some geologists and biologists to imagine great, elongated isthmian links between continents. After all, the Isthmus of Panama is a good modern example. If North and South America are today joined by a long, narrow land bridge, why could not other continents have been similarly linked in the past? Why could there not have been a long land bridge from Africa to Antarctica, affording a dry-land passage that permitted *Lystrosaurus* and other reptiles and amphibians to reach the south polar continent? Theoretically such might have been the case; actually there is no geological evidence for such a bridge. Upon the basis of our modern understanding of earth structures and earth forces, there are no indications that such an isthmian link once existed or that it is now sunk into the depths of the ocean.

This brings us, therefore, to the alternative hypothesis, namely that in early Triassic time (and probably before then, as well) Antarctica was closely joined to southern Africa; that the two continents were indeed contiguous and a part of the same land mass. The fact that *Lystrosaurus* in Antarctica is morphologically so very similar to *Lystrosaurus* in Africa points strongly to this possibility. In short, the fossils of *Lystrosaurus* from Africa and from Antarctica have the appearance of being the remains of animals that roamed freely across a common ground; they do not look like the bones of animals that inhabited two separate regions, with a narrow connection between the two areas. *Lystrosaurus* in Antarctica and in Africa would appear to be, quite possibly, the representatives of a single species inhabiting a single range.

This is strong evidence indeed for a broad ligation between Antarctica and Africa at the beginning of Mesozoic history. And it is probably not in the least fortuitous that the evidence of *Lystrosaurus* and other fossil vertebrates is supported by various other lines of paleontological evidence, such as that offered by the distributions of fossil plants and certain invertebrates, as well as by geographical and geological facts of the most convincing nature.

The relationships of ancient continents to each other frequently have been interpreted according to the relationships between modern continents, particularly by their outlines. As we have already seen, the correspondences between the shores of South America and Africa have long attracted the attention of geographers and geologists, and were instrumental in Wegener's early thinking about ancient continental relationships. Can such correspondences be entirely accidental? The comparisons that have been made through the years by simple inspections of increasingly accurate maps and globes would seem to indicate that there is more than mere chance involved in the resemblances of the continental edges on the two sides of the South Atlantic. But the eye can be fooled, even under the best of circumstances.

In recent years, however, some geologists and geophysicists, notably Professor Edward Bullard of England and others who have followed his lead, have made sophisticated, computerized studies of the continents, basing such studies not upon the strands, where land meets water, but upon deep contours, at about the 500- or even 1,000-fathom lines, where the transition from the shallow continental shelf to the deep ocean floor occurs, and where it may be presumed one finds preserved in some degree the outlines of the ancient continents. The results as applied to Africa and South America are most interesting, as we shall see.

But at the moment let us devote our attention to Antarctica and Africa, where this method of comparing continental outlines has been applied quite recently at the 500-fathom line, or isobath, by A. G. Smith and Anthony Hallam, and at the 1,000-fathom isobath by Robert S. Dietz and Walter Sproll. Both of these independently conducted studies have yielded similar results; both show that there is a remark-

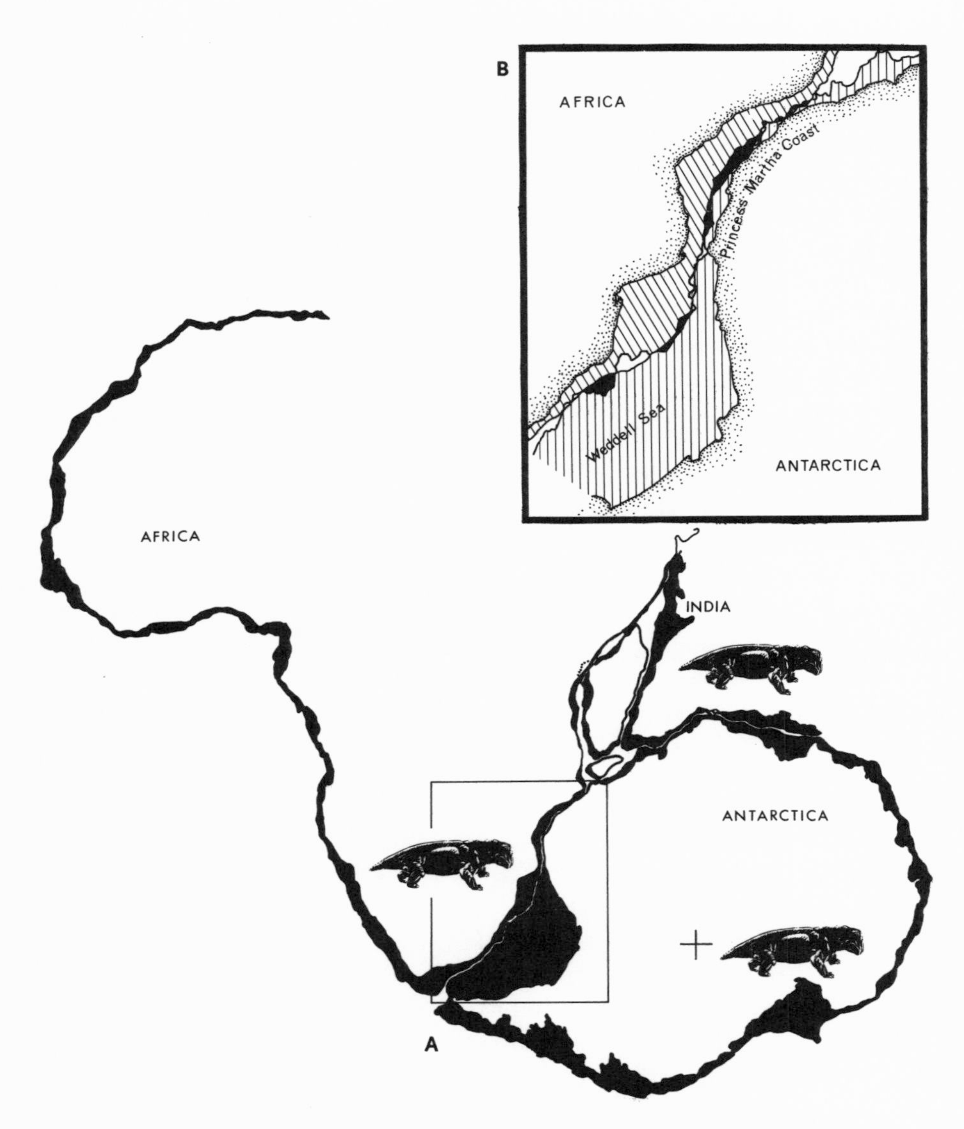

FIGURE 26. Evidence for Gondwanaland. A. Africa, Madagascar, peninsular India, and Antarctica fitted together at the 1,000-fathom isobath (6,000 feet below sea level), shown here by the black areas beyond the present coastlines. The drawings of *Lystrosaurus* are located approximately where this reptile and associated vertebrates are found in Lower Triassic sediments on the three continents. This illustration emphasizes the remarkable fit between the continents and the integration of *Lystrosaurus* distributions (now widely separated) within the limits of a possible range for a single tetrapod fauna. B. An enlarged map of the fit between Africa, from Mozambique to Natal, and Antarctica, from the Weddell Sea to the Princess Martha Coast, at the 1,000-fathom isobath. Black areas indicate overlaps; blank areas indicate gaps.

able correspondence between the southeastern margin of the African continent, from the general region of Durban, Natal, to the northern portion of Mozambique, and the western border of Antarctica, from the Weddell Sea to the Princess Martha Coast. When the two continents are shifted so that the above-mentioned borders come together, the fit between them is nothing short of remarkable. They join each other like pieces in a jigsaw puzzle, with only slight gaps and slight overlaps here and there. The correspondence over a distance of 1,500 miles or more is so close that it stretches the imagination almost to the breaking point to think that it is all a matter of coincidence. Here, it would seem, we are looking at the two sides of a great sinuous fracture, which at one time in the distant past marked the separation of the Antarctic land mass from its African counterpart.

It may be remembered that in the previous chapter something was said about the general geological similarity between the Permo-Triassic rocks of Antarctica and those of southern Africa, extending from perhaps 250 million years ago to about 200 million years in the past. In both continents there are great thicknesses of flat-lying sediments, representing the later phases of Permian history and much of the Triassic. In both continents these sediments are intruded by heat-formed rocks known as "dolerites," and frequently capped by massive flows of volcanic rocks known as basalts. And in both continents the sediments of earliest Triassic age contain, of course, *Lystrosaurus* with associated fossils.

THE SAMFRAU GEOSYNCLINE, OR SOMETHING LIKE IT

The resemblances go beyond this. For example, the Permo-Triassic rocks containing *Lystrosaurus* and other fossils in the south polar continent are exposed in the steep and frequently terrifying cliffs of the Transantarctic mountains, a stupendous range that crosses Antarctica from Victoria Land to the Weddell Sea, its peaks rising 12,000 to 14,000 feet and more into the Antarctic sky. Paralleling this mountain range are the Precambrian and Paleozoic geosynclines of Antarctica (geosynclines being elongated belts of complexly folded rocks) derived from ancient troughs of accumulated sediments on a grand scale. If this does not convey a mental picture of a geosyncline, think of a thick book, the pages of which have been crumpled by being pushed from their free edges toward the gutter of the binding. Or think of the effect produced by squeezing, from two sides, a thick, plastic block composed of

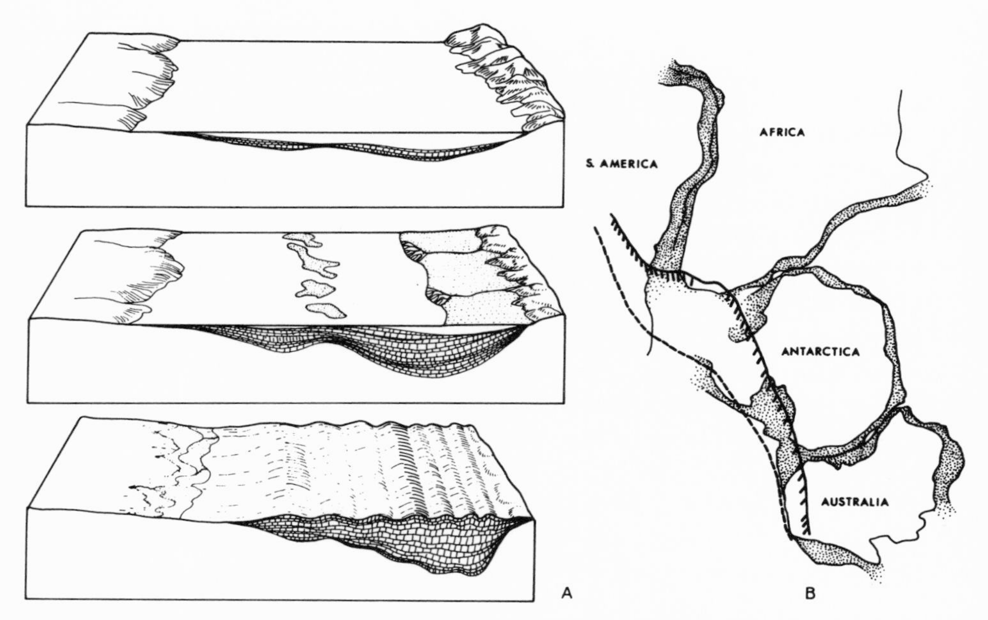

FIGURE 27. A. The block diagrams show some stages in the development from a geosyncline to a belt of folded mountains. The top figure shows the accumulation of sediments derived from bordering land masses, in relatively shallow waters. In the middle figure the geosyncline is subsiding, owing to the weight of the accumulating sediments. In the bottom figure the thick sediments of the geosyncline have been folded by powerful earth forces. B. The geosynclinal belt, called by Du Toit the Samfrau Geosyncline, that reaches from South America, across the tip of Africa, and thence through Antarctica and Australia. The alignment of the geosynclinal structures on the several continents, when the continental blocks are placed in the positions shown, constitutes a strong argument for an ancient supercontinent of Gondwanaland.

a series of differently colored layers of soft clay. And then magnify such a model to the dimensions of mountain ranges and valleys. These same Antarctic geosynclines may be aligned with the Adelaide and perhaps the Tasman geosynclines of Australia, when this latter continent is placed so that the Great Australian Bight, forming the southern border of the continent, is fitted against Wilkes Land in Antarctica. And again, when Africa and South America are brought together, the folded Cape mountains are in alignment with the Sierra de la Ventana of Argentina, the two systems of folded mountains showing extraordinarily strong resemblances to each other.

This alignment of geosynclinal structures from South America seemingly through the southern tip of Africa, across Antarctica, and thence into Australia was first projected more than 40 years ago by the brilliant South African geologist A. L. Du Toit, to whom we have already been introduced as one of the pioneers in the study of continental drift. He named it the "Samfrau Geosyncline." Perhaps this hypothetical geosyncline today is not universally accepted in its entirety; although the resemblances between the Cape Folded Belt and the Sierra de la Ventana are remarkably close, and although the geosynclines of Antarctica may be correlated quite satisfactorily with those of Australia, there is some question as to whether these structures in their totality may logically be joined into a single system. Recent work, however, does indicate that there was probably a geosynclinal system similar to, if not identical with, Du Toit's Samfrau Geosyncline, which crossed the several continents in late Paleozoic and early Mesozoic times. This aligned series of folded rocks—the Gondwanide Fold Belt, as it may be called—is a structural feature linking the several parts of Gondwanaland, and controlling their association together, as is corroborated by the juxtaposition of matching continental borders. Indeed, the alignment of the Gondwanide Fold Belt is probably the key to the reconstruction of Gondwanaland; any juxtaposition of the present continents to form Gondwanaland must be in accord with the Gondwanide Fold Belt alignment, to be valid. Thus the large-scale study of geologic structures would seem to accord with the evidence of the fossils, as well as the evidence of continental outlines, to indicate a close connection between Africa and Antarctica in the years of *Lystrosaurus*.

PALEOMAGNETISM

The study of paleomagnetism has, in recent years, pointed to interesting possibilities for reconstructing the ancient positions of continents. Simply expressed, paleomagnetism is "fossilized magnetism" preserved in ancient rocks. It is known that when volcanic rocks cool from their original molten state, the magnetic particles within the rock are aligned in relation to the magnetic poles of the earth. This expression of magnetism is then "frozen" into the rocks and remains, no matter how the rocks may be turned or twisted by subsequent geological events. Therefore, by comparing the alignment of magnetism of certain volcanic rocks now found in various continents (taking into account, of course, the "turning" and "twisting" the rock has been subject to), it is possible to recreate the attitudes of those continents in relation to the positions of the magnetic poles (as they then were placed) at the times when the rocks were formed. It is a scientific method fraught with many problems, and it rests upon various assumptions (such as the one that the magnetic poles in past ages were reasonably close to the poles of the earth's rotation), but when applied circumspectly it would seem to give us considerable insight as to continental relationships during the extent of geological history. And such studies would again seem to demonstrate a close relationship between Antarctica and Africa.

GONDWANALAND—THE PHYSICAL EVIDENCE

So much for some of the particulars of continental drift. At the moment we are concerned with Gondwanaland, the postulated southern-hemisphere supercontinent that was ancestor to the modern continents located largely to the south of the equator, as well as peninsular India.

What is the evidence for this ancient southern supercontinent?

The evidence began, as we have seen, by a comparison of the eastern coastline of South America with the western coastline of Africa. This comparison has been immeasurably strengthened in recent years by the use of computers, as we have seen in the comparison of Africa and Antarctica. For readers who wish to know something of the method in terms somewhat more precise than those of this general discussion, the following quotation is given:

> The method uses a computer program, written by J. E. Everett, to determine the best least-squares fit of two lines on a sphere. The lines to be fitted together are digitized at intervals of roughly 50 km (about half a degree of latitude). A "homing" routine systematically searches for the best fit of the lines by rotating one of them about a movable point (the rotation pole) until it fits the other line as well as possible. (Smith, A. Gilbert, and A. Hallam, *Nature*, 1970, Vol. 225, p. 139.)

No further attempt will be made here to explain the method; certainly it is justified by the results. For more than 4,000 miles along their borders the two continents fit along the thousand isobath fathom in a manner that is truly remarkable. Here and there are some overlaps, here and there are some gaps. Yet all in all the correspondence is close beyond any possibility of coincidence.

So it is that in addition to the fit between Antarctica and Africa, there is this remarkable fit between Africa and South America—the first to be recognized in the study of past continental relationships. Yet these are not the only close fits between southern continents. The placing of the Great Australian Bight against Wilkes Land in Antarctica has already been mentioned. This, too, is a fit that can be justified by computerized studies and one that, like the other fits, is too perfect to be a coincidental resemblance. Similarly, the peninsula of India, Ceylon, and Madagascar can be fitted between Antarctica and Africa, with the eastern border of India in juxtaposition to Enderby Land in Antarctica, with its western border against the eastern border of Madagascar to the south and Somalia to the north, and with the western border of Madagascar opposite Tanzania and Kenya. Ceylon presumably occupied the relatively small area between the southern tip of the island of Madagascar (the Malagasy Republic), the northern portion of Mozambique, and Queen Maud Land in Antarctica.

It has recently been suggested, upon the basis of geophysical evidence, that most of China, and perhaps Indochina, may have been a part of Gondwanaland, forming a northeastern extension of the ancient continent, to occupy much of the area between Africa and Australia. Such is one interpretation of the composition of Gondwanaland, according to the theory of continental drift. The relationships of its several component parts are perhaps made clear by the map on page 118.

This fitting together of an ancient jigsaw puzzle is convincing, to say the least. One fit might be explained away, although the one between South America and Africa would require a considerable amount of explaining. But a half-dozen fits, all of them remarkably close and all involving continental borders measuring in each case thousands of miles in length, are too much to attribute to coincidence.

The reconstruction of Gondwanaland upon the basis of continental outlines is not the only evidence for favoring the reality of this ancient continent. As we have seen, in the comparison of Africa and Antarctica, there are other resemblances involving geological and paleontological comparisons that accord with the evidence of continental outlines. Let us extend the comparisons that already have been made between Antarctica and Africa to the whole of Gondwanaland.

In the comparison of Antarctica and southern Africa some mention was made of the folded geosynclinal structures that are present in the Antarctic continent and Australia, as well as in South America and the southern tip of Africa. It will be recalled that these folded belts are brought into alignment when the southern-hemisphere continents are grouped in the manner already described.

Certainly the presence in Antarctica and South Africa of great thicknesses of Permo-Triassic sediments of terrestrial origin, these cut

FIGURE 28. South Africa. Horizontal Permo-Triassic sediments intruded by volcanic rocks—the characteristic scenery of the Karroo basin.

and intruded by dolerites and capped by extensive volcanic basalts of Jurassic age, forms cumulative evidence for the close proximity and essential unity of these two continental regions during late Paleozoic and early Mesozoic history. The comparison may be extended to include southern Australia, where the dolerites of Tasmania, so impressively displayed in Mount Wellington, near Hobart, are of the same geological age and have the same unusual chemical composition as those of Antarctica.

The presence of some ancient Precambrian rocks known as "anorthosites" may cast additional geological-mineralogical light on the former disposition of the southern continents. These anorthosites occur in South America to the north of the Rio de la Plata, across Africa, from southwest Africa to Mozambique, across Madagascar, through the eastern portion of peninsular India, and in central Australia. With Gondwanaland constituted in the way that is here being advocated, the ancient anorthosites take the form of a long, curved belt, sweeping through the ancient supercontinent to the north of the Gondwanide Fold Belt, but, like this last-named structure, geologically binding the supposed former land mass together.

FIGURE 29. Antarctica. Horizontal Permo-Triassic sediments intruded by volcanic rocks (and largely covered by snow)—the characteristic scenery of the Transantarctic mountains.

We have already seen how paleomagnetism in the rocks would seem to link Antarctica with Africa. By extension, this evidence applies to the whole of Gondwanaland. Thus, paleomagnetism as preserved in Permian and Triassic rocks in Africa, for example, indicates that the poles during those geologic ages were located at certain positions far removed from the poles of today. Similar evidence from South America gives quite different locations for the poles. And the same holds for Australia—and so on. But when the various positions for the poles, as indicated by such evidence, are brought together, or at least in close proximity, the continents become closely associated in patterns essentially the same as those indicated by the evidence of continental outlines and geological features. So once again there is corroborative evidence for the continent of Gondwanaland.

Still another line of evidence that points with particular force to the probable existence of a Gondwana continent is the presence of Permo-Carboniferous tillites (which are hardened glacial deposits), associated with glacially striated rocks, in South Africa, Brazil, Madagascar, peninsular India, and Victoria, Australia. With the continents in their present locations the occurrences and attitudes of these Indian and southern-hemisphere tillites and striations show an illogically haphazard arrangement. Moreover, as the continents are now situated, the glacial striations are variously directed. Those in South Africa are oriented toward the south or the southwest. Those in South America show an east to west trend. And those in India and Australia are generally northward in direction.

Furthermore, the Indian and South American tillites are in tropical or subtropical regions—which might be possible if the glacial evidences were locally limited, in other words, if they represented the work of glaciers flowing from high mountain peaks. But the tillites and striations are of wide extent, not only in these continents, but also in South Africa and Australia. They are obviously the record of a widespread glaciation.

If the southern continents and peninsular India are grouped into a continent of Gondwanaland, according to the pattern that has been set forth in some of the preceding discussion, the Permo-Carboniferous tillites and striations become associated in a manner that is logical and understandable. Instead of being scattered through more than 200 degrees of longitude, the tillites are arranged roughly in a circle, surrounding Antarctica. And with this arrangement it is especially significant that the glacial striations show a radiating pattern, extending outwardly from a center that would seem to have been to the east of South Africa—and possibly within the confines of east Antarctica.

FIGURE 30. The evidence of Permo-Carboniferous glaciation in Gondwanaland. The arrows indicate the directions of glacial striations on Permo-Carboniferous rocks. The circles show the presence of presumed tillites, which are glacial deposits. The pattern indicates what may have been the extent of an ancient ice cap.

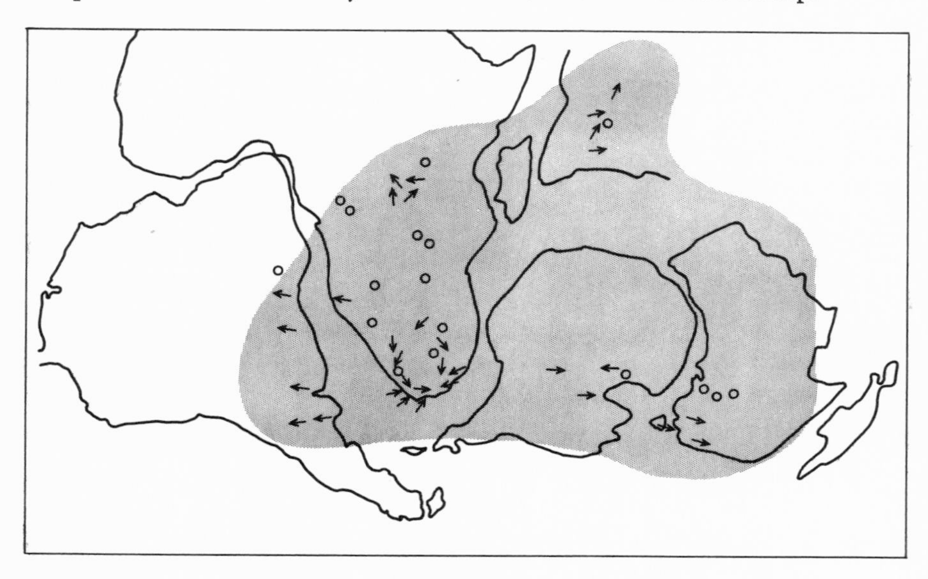

All of this evidence would seem to indicate, in a most positive fashion, that the Permo-Carboniferous tillites and striations of the southern continents and peninsular India are the record of a massive ice cap that covered a large portion of Gondwanaland during a part of late Paleozoic time. All of this evidence constitutes a remarkably strong argument in favor of Gondwanaland.

Thus far we have been looking at the physical criteria for an ancient Gondwanaland, consisting of several lines of evidence, each of which reinforces the other. Let us now turn our attention to some of the paleontological evidence that bears upon this problem.

GONDWANALAND—*MESOSAURUS* AND *GLOSSOPTERIS*

Wegener and Du Toit both attached a great deal of importance to the Lower Permian reptile *Mesosaurus*, designated by the latter author as "that unique reptilian genus *Mesosaurus* from the Dwyka shales of the Cape and the Iraty shales of Brazil, Uruguay, and Paraguay." *Mesosaurus* is indeed unique—in more ways than one. It is the single genus representative of an entire order of reptiles, the Mesosauria (although some authorities would place the South American fossils within a separate genus, *Stereosternum*, a procedure that probably is not justifiable). The Mesosauria are found only in two general areas of the world, South Africa and South America, in the regions mentioned above. Therefore Wegener and Du Toit argued that *Mesosaurus* constitutes strong evidence for the former junction of the two continents now bordering the South Atlantic ocean.

But there have been questions about *Mesosaurus* in the past—questions that now are resolved upon the basis of other fossil reptiles, but questions which, before such resolution, were legitimate and troublesome. The problem is that *Mesosaurus* was quite obviously an aquatic reptile, swimming freely and rapidly through the water in pursuit of fishes, a conclusion plainly indicated by the nature of the skeleton of this little fossil. Thus the skeleton of *Mesosaurus* is elongated and supple, being something on the order of a foot and a half in length. The tail is flexible and obviously was an efficient swimming organ, while the limbs are modified as large paddles. The skull is very much elongated by a prolongation of the tooth-bearing portions of the upper and lower jaws, and these are set with numerous very long, slender teeth, these several adaptations being quite patently constructed for the capture of small fishes. The nostrils are set far back, just in front of the large eyes, an adaptation that occurs time and again in air-breathing aquatic vertebrates.

If *Mesosaurus* were so nicely specialized for life in the water, then why would a connection between Africa and South America be necessary to explain the presence of this little reptile in these two continents and in no other Permian sediments throughout the world? (The absence of *Mesosaurus* on other continents might be passed off as owing to the accidents of preservation and collecting, but this seems hardly likely today, in view of our rather extended and worldwide knowledge of fossil deposits.) Could not *Mesosaurus* have swum from the one continent to the other? Probably not, because the testimony of the sediments in which this little reptile is preserved indicates that it was a fresh-water or brackish-water animal. Nevertheless, the possibility remains. We know that various modern crocodilians, which habitually live in rivers and lakes, do venture into oceanic waters. Probably *Mesosaurus* was not capable of "breasting the waves over the several thousand miles of oceans which now separate these two continents [Africa and South America]," as Professor Romer has stated. Nevertheless, if there had been any kind of an island arc stretching between what many authorities considered as two separate continents during early Permian time, then it is conceivable that *Mesosaurus* might have crossed from the one region to the other by such a route.

So, until the discoveries of virtually identical land-living reptiles (of Triassic age) in South America and Africa, within recent years, the position of *Mesosaurus* as a paleontological link between the two southern-hemisphere continents in which its fossils are found was equivocal, and quite rightly so. Now, as we shall see, the southern-hemisphere occurrences of other fossil reptiles make the distribution of *Mesosaurus* within a single continental area more plausible than it once was. Thus we are justified in regarding *Mesosaurus* probably as good paleontological evidence for the union of South America and Africa during early Permian time.

There is a dearth of other fossil vertebrates of Permian age in the South American continent; consequently no adequate base exists for comparisons with the remarkably rich Upper Permian reptilian faunas of South Africa. But fossil evidence of Permian age that has been widely cited as an indication for the former existence of a continent of Gondwanaland is found in the plant *Glossopteris,* widely distributed across the continents of the southern hemisphere. *Glossopteris* was a seed fern with large, spatulate leaves, and so typical is this plant of Permian sediments in various southern continents that its areas of distribution are often referred to as coming within the limits of the "*Glossopteris* flora." Fossils are found in South Africa, South America, India, Australia, and Antarctica, in the latter within about 300 miles of the South Pole. Such a distribution, it is argued, is evidence for the former inclusion of the several continental regions just listed within the single continent of Gondwanaland. But again, as in the case of *Mesosaurus*, there has been in former years some doubt as to the meaning of this distribution. Could not the seeds of *Glossopteris* have been blown by strong winds across water barriers from one land mass to another? Many paleobotanists have denied this possibility with vigor, yet the doubt remained. But again, as in the case of the evidence of *Mesosaurus*, the discoveries in recent years of land-living vertebrates that could not possibly have crossed wide oceanic barriers, nor could they have been blown by winds, have given weight to the distribution of the *Glossopteris* flora as plausible evidence for the former existence of Gondwanaland.

THE ANCIENT REPTILES OF GONDWANALAND

This new evidence from land-living vertebrates comes from the Triassic rather than the older Permian. First and foremost among the animals constituting such evidence is our old friend *Lystrosaurus*. By now the reader probably is sick unto the point of ennui of *Lystrosaurus*, yet there is no getting away from this useful reptile. With the evidence of *Lystrosaurus* at hand to show that certain southern continents almost indubitably were connected in early Triassic times, the presence of various other reptiles of that age on the two sides of the South Atlantic can now be accepted as further indications of close land connections, without some of the doubts that previously had existed. The idea of a "long way around" between South America and Africa, a route that would involve intercontinental migrations through northern Africa, Asia Minor, Asia, across the Bering region, down through North America, across an isthmian bridge, and thus to South America, irrevocably fades into the realm of unreality.

Those various reptiles that span the South Atlantic, mentioned in the foregoing discussion, are somewhat higher (or later) in the Triassic sequence than *Lystrosaurus*. *Lystrosaurus*, so typical of the earliest Triassic sediments of South Africa and Antarctica, as well as of peninsular India and western China, has not as yet been found in

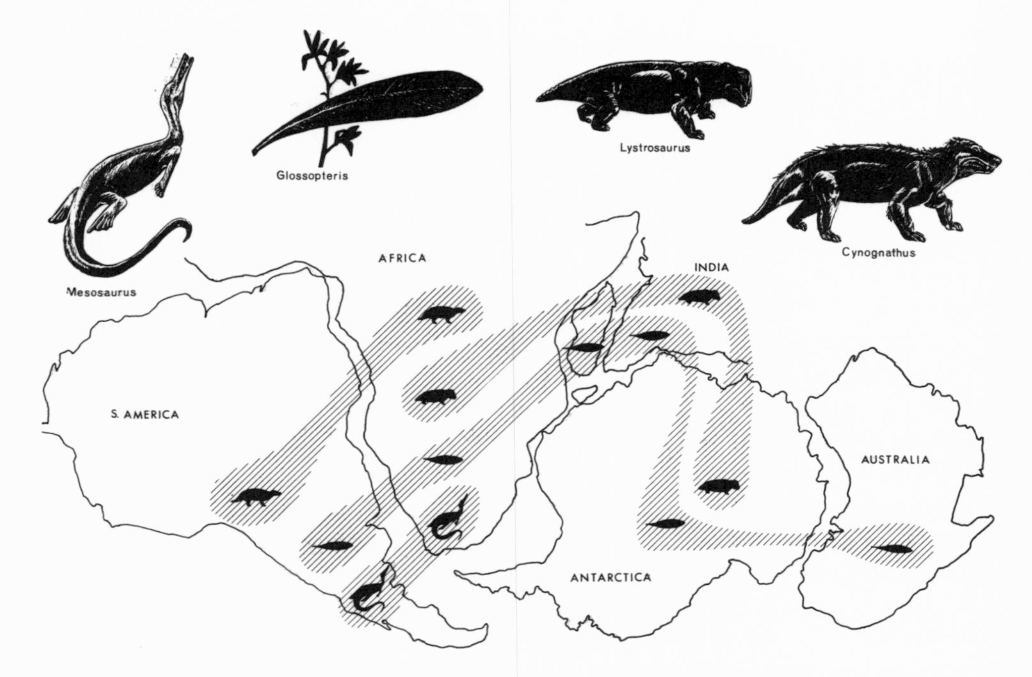

FIGURE 31. Gondwanaland reassembled, and some of the paleontological links that bind it together. *Mesosaurus,* a Permian reptile, occurs in southern Brazil and South Africa; *Glossopteris,* a Permian plant, occurs across all of the Gondwanaland components; *Lystrosaurus,* a Lower Triassic reptile, occurs in South Africa, peninsular India, and Antarctica (and in southeast Asia as well); *Cynognathus,* a Lower Triassic reptile (somewhat later in age than *Lystrosaurus*), occurs in Argentina and South Africa.

South America. Perhaps it will be some day; there would seem to be no reason why it should not occur there. But at the present time we must base our conclusions on the materials at hand, and in South America there is no *Lystrosaurus* at hand.

There is, however, a very progressive mammallike reptile known as *Cynognathus.* This Triassic reptile, a specialized predator, was one of the more advanced of the mammallike reptiles. The skull was very doglike (hence the name, cyno = dog, gnathus = jaws) and the teeth were nicely adjusted to a carnivorous diet. There were small nipping teeth in the front of the jaws, large, piercing canines on each side, and behind the canines, bladelike cheek teeth, each with several cusps, well-adapted for tearing and cutting flesh. The body was slender and supple; the legs and feet were somewhat elongated, and obviously suited for rather rapid running. *Cynognathus* may be thought of as a sort of reptilian "wolf," and indeed some of the fossils of this ancient hunter indicate an animal as large as a wolf. Generally speaking, however, most specimens of *Cynognathus* are of the size of small foxes. This interesting carnivore, obviously an aggressive predator probably feeding upon other reptiles, is characteristic of the sediments that constitute the *Cynognathus* zone, immediately above the *Lystrosaurus* zone, in the South African sequence. *Cynognathus* in South America is found in western Argentina, in the Puesto Viejo Formation. Various other fossil reptiles occur in association with it, of which particular mention may be made of another carnivore, *Pascualgnathus*, and a large dicynodont, a tusked reptile related to and looking something like *Lystrosaurus,* and known as *Kannemeyeria. Kannemeyeria* is also very characteristic of the *Cynognathus* zone in South Africa, as are reptiles closely related to *Pascualgnathus*. Thus there are multiple links among the fossils between the two continental regions, and since these fossils represent highly specialized terrestrial animals, animals that spent their lives in active movement over dry uplands, we have strong evidence at hand to indicate the close connection between South Africa and South America. In fact, *Cynognathus* and *Kannemeyeria* in the two regions are so very closely comparable that it seems reasonable to think that they are representatives of animals occupying a single, large area of distribution, just as *Lystrosaurus* probably indicates this same fact for South Africa, Antarctica, and peninsular India, in the time span at the beginning of Triassic history, preceding the development of the *Cynognathus* fauna.

It may be recalled that in India *Lystrosaurus* occurs in the Panchet beds, which represent the onset of the Triassic record in that continent. Above the Panchet beds is the Yerrapalli Formation, and recently there has been described from the Yerrapalli sediments a dicynodont reptile, related to, but somewhat different from the South African

FIGURE 32. The skulls of two large Triassic dicynodont reptiles: *Kannemeyeria argentina* from Argentina (left) and *Kannemeyeria erithrea* from South Africa (right). An example of the close paleontological ties between two components of Gondwanaland.

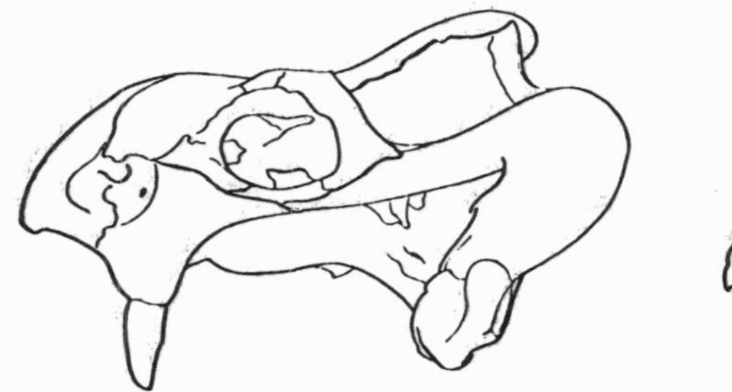

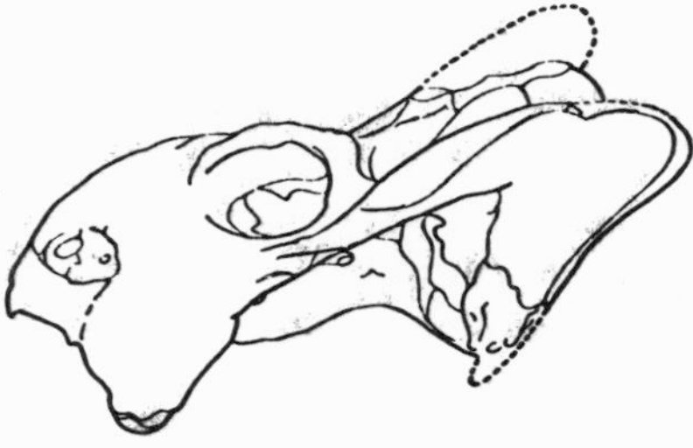

and South American *Kannemeyeria.* This newly described kannemeyerid (a term indicative of its general relationships) is one of a series of related dicynodont, or two-tusked, reptiles found not only in the Indian portion of Asia, but in western and central China as well, to indicate that these strange reptiles roamed in great numbers and considerable variety across early Triassic lands.

So far we have been concerned with comparisons between land-living reptiles of early Triassic age. Resemblances are evident, however, when we look at later Triassic reptiles. For example, in the Middle or Upper Triassic Ischigualasto and Santa Maria formations of Argentina and Brazil respectively, are found several dicynodont reptiles closely related to dicynodonts in Africa and India. There also are some very interesting specialized carnivorous reptiles related to *Cynognathus.* These are cynodont reptiles which, although in many respects resembling *Cynognathus,* are distinctive in having very broad cheek teeth—teeth that resemble the broad molars of many mammals, bears for example. It should be understood that no direct relationship is implied; perhaps, however, these particular reptiles, which are often called "gomphodonts" and among which there are several varied genera, were omnivorous carnivores, subsisting on a diet of meat and plant food, as do bears, raccoons, and skunks. Closely related broad-toothed gomphodonts are found in Africa.

To get away from the mammallike reptiles, it might be helpful to consider briefly some very strange later Triassic reptiles known as "rhynchosaurs." These reptiles, peculiar to certain Triassic faunas, were the distant and very large cousins of the modern tuatara, *Sphenodon,* of New Zealand. In appearance, however, the rhynchosaurs had little in common with the rather lizardlike tuatara. For the rhynchosaurs were massive animals, often as large as sheep, or even larger, with heavy bodies, short limbs, and short tails. The skull was very peculiar, being short, deep, and wide, with two terminal "tusks" in front, one on each side of the midline. But these were not tusks in the ordinary sense of the word; for some strange reason they are the premaxillary bones of the skull, developed to perform the function of tusks. What might have been the adaptive advantage of such an arrangement (for exposed living bone is quite susceptible to infection) is hard to imagine. However that may be, such was the adaptation characteristic of the rhynchosaurs. In the sides of the jaws are peculiar batteries of small, knoblike teeth.

It is a matter of conjecture as to how the rhynchosaurs lived or

FIGURE 33. A restoration of a rhynchosaur. These strange reptiles were especially characteristic of Gondwanaland, particularly South America, Africa, and India, during middle and late Triassic times.

what they ate. They are found, for example, in the Santa Maria Formation of Brazil, but they are *never* found in association with the dicynodonts in that formation. Were the rhynchosaurs and dicynodonts incompatible? It would appear that such might be the case. Perhaps they lived in different environments. Perhaps the rhynchosaurs ate husked fruits of some sort that determined the habitats in which they lived.

Wherever they lived, they certainly inhabited these regions in great abundance. The fossils of rhynchosaurs, which occur in profusion in the Santa Maria Formation of Brazil near the city of Santa Maria, Rio Grande do Sul, are also found in the Ischigualasto Formation of Argentina. Similar rhynchosaurs occur in the Middle Triassic Manda beds of east Africa, and in India, in the Upper Triassic Maleri Formation of the peninsula. They would seem to indicate that the physical conditions that allowed land-living reptiles to become dispersed between different portions of Gondwanaland continued at least into the final stages of Triassic history.

EARLY AMPHIBIANS OF GONDWANALAND

So far the discussion of land-living vertebrates in Gondwanaland has been concerned with reptiles. What about the amphibians?

The first fossil vertebrate of significant geological age to be discovered in Antarctica was a fragment of a labyrinthodont amphibian jaw, as was described in the first chapter of this book. Subsequent work in Antarctica has revealed more amphibians, some large and some small. And these would appear to be related to large and small amphibians found in the *Lystrosaurus* zone of South Africa. Moreover, some of the Antarctic amphibians, especially the smaller ones, show resemblances to Lower Triassic labyrinthodont amphibians that have been found in the last few years in Australia. This is an important comparison.

The Triassic sediments of Australia are of such origin that they contain amphibian bones to an overwhelming degree; there are very few reptiles known. This is, of course, a reflection of the environments in which the sediments accumulated, perhaps low swamps and ponds, rather than rolling uplands traversed by numerous streams, with interspersed lakes. The Lower Triassic amphibians of Australia, found particularly in Tasmania and in northwest Australia, are, as mentioned, related to amphibians found in Antarctica and to amphibians from the *Lystrosaurus* zone of Africa. Thus we see fossil evidences for a connection, to back up the physical evidences that already have been cited, between Australia and Antarctica, and evidently through Antarctica to South Africa.

To this point our concern has been with the fossil evidence as it bears upon the possible existence of a Gondwanaland. And to this point our concern has been restricted to Upper Paleozoic and Lower Mesozoic fossils, representing animals that lived 200 million years and more ago. Do the fossils of later ages tell us anything? They do, but they indicate continental relationships evidently changed from those which held during the times with which we have been occupied. So they will be discussed in subsequent chapters.

THE BEECH FORESTS OF THE SOUTHERN HEMISPHERE

Do the distributions of modern plants and animals give us clues concerning Gondwanaland? Many authorities have insisted that they do, and they cite numerous examples, such as the southern-hemisphere beech, *Nothofagus,* now found in South America, Australia, New Zealand, and New Caledonia, or the lungfishes of Africa, South America, and Australia. Suffice it to say here that the southern-hemisphere beeches, at least in part, may very well represent wind-borne dispersals, and the lungfishes are remnants of fishes that were of worldwide distribution in Triassic times. There are other examples of modern plants and animals that have been used to show how they may represent remnants from an ancient Gondwanaland. They will not be listed here. Some of these distributions may in fact be relics from a former Gondwanaland, but others may be the results of different and more recent modes of spreading from one continental region to another. There are too many complex factors of animal and plant distribution—the result of earth history since Permo-Triassic times—for modern organisms to be interpreted as indications of ancient Gondwanaland relationships, except with the utmost circumspection. Darlington, whose wide-ranging studies in the biogeography of both modern and extinct organisms are noteworthy by reason of the thoroughness and careful interpretations with which they have been made, has suggested that "plant and animal distributions as now known do not show where the earlier connections were. Only in the late Cretaceous and especially in the Tertiary do plant and animal distributions begin to show specific land connections and specific ocean barriers, and this is too late to be significant in any likely hypothesis of continental drift" (Darlington, 1965, p. 197). Such being the case, our attentions will be directed, as they have been, to the fossil forms.

By now it is abundantly established that there is a large body of evidence indicating the possible former existence of a supercontinent of Gondwanaland in the southern hemisphere. This evidence is, as we have seen, in part physical and in part paleontological. And tiresome as may have been the foregoing recital of numerous facts and their interpretations, the list is far from being exhausted. Perhaps, however, it has been sufficient to make the point—which is that modern studies in many fields have yielded results all indicating the probable reality of Gondwanaland in late Paleozoic and early Mesozoic times. What about the rest of the world, which is to say the northern hemisphere? One cannot discuss Gondwanaland and ignore the land masses that did not participate in the formation of that great southern continent.

THE COMPOSITION OF LAURASIA

The corollary to Gondwanaland is *Laurasia,* the hypothetical northern supercontinent made up of what are now North America, Greenland, and Eurasia. Gondwanaland has been given first consideration, because the concept of Gondwanaland was the genesis of the larger idea of an earlier earth markedly different in the relationships of land masses from the earth with which we are acquainted. With the development of the idea of an ancient Gondwanaland, the idea of a northern counterpart to the great southern supercontinent logically followed.

The same techniques can be applied to the northern-hemisphere continents that were used in the study of continental relationships south of the equator; they have been used, and the results are strikingly parallel to those obtained in the reconstruction of Gondwanaland. Since there has been a rather full discussion of some of the methods used and the conclusions drawn in arriving at our current ideas as to the composition of Gondwanaland, it is not necessary to repeat the arguments in detail as they apply to Laurasia. Nevertheless some summary statements are in order.

First consideration is again given to the problem of continental outlines. When the edges of the northern continents at the 500-fathom line are tested by computer, a series of remarkable fits ensues. It must be remembered that continental shelves are rather extensive off the eastern coast of North America and the western coast of Europe, so that the fit of these continental masses is not at first obvious, especially if one regards only the shorelines with which we are familiar. But at the 500-fathom line—the depth used by Sir Edward Bullard and his associates in these studies—the fits between continents are essentially as convincing as those between the southern-hemisphere land masses.

Thus the northwestern border of Africa, from Mauretania to Morocco, abuts nicely against the continental margin of North America, from Florida to Newfoundland, with overlaps and gaps of relatively small dimensions. The southern border of Spain fits against the Moroccan-Algerian edge of northern Africa. To the north, the western border of Greenland fits against Labrador and Baffin Island, with the Davis Strait essentially closed, while the eastern border of the Greenland block joins the north European continental block, from England and Ireland to the northern part of the Scandinavian peninsula. Arabia may be considered as an integral part of the African block, joined to that part of Africa to which it is now opposite, and separated from it only by the narrow Red Sea. To the east the land mass of Laurasia was continued by what is now eastern Europe, and perhaps all of Asia north of the Himalayan region.* Such a reconstruction of Laurasia envisages a great east-to-west supercontinent in the northern hemisphere, stretching across some 180 degrees of latitude and separated in its eastern section from Gondwanaland by an east-to-west sea, which has been called the "Tethys Sea."

This composition of the Laurasian land mass is indicated not only

* As mentioned on page 65, some recent studies suggest that most of China may have belonged to Gondwanaland, rather than to Laurasia.

by the remarkable fitting together of the continental blocks, but also by various geological criteria of the sort that support the reconstruction of Gondwanaland. For instance, the paleomagnetic evidence for the combining of the northern continents into a Permo-Triassic Laurasia is strong. Moreover, there are geological features on the two sides of the North Atlantic that support the former juncture of the continents that now border this ocean, with the consequent absence of any oceanic barrier at the time the continents were joined. Thus certain great faults (dislocations caused by the slipping of rock masses along fracture planes), which cut across the British Isles to form prominent features (glens and lochs) in Scotland and Ireland, would appear to be continuations of faults that cut across Newfoundland. The faults and their displacements of early and middle Paleozoic sediments are quite comparable in the two regions, now separated by the Atlantic Ocean. We seem to be looking at portions of what was once a single continent.

If Laurasia is so reconstructed, with the North Atlantic ocean as yet unformed in Paleozoic and early Mesozoic times, the relationships of some of the fossil vertebrates of North America and Eurasia, once a thorny problem for paleontologists, are rather satisfactorily resolved. In Czechoslovakia, for example, there are Carboniferous shales more than 300 million years old, containing the numerous remains of early amphibians, especially labyrinthodonts, and very primitive reptiles. The fossils from these Middle European sediments show many close relationships to Carboniferous amphibians and reptiles found in eastern North America. On the conventional globe these fossil deposits are 5,000 miles apart, with no communication between them except by a long route of 15,000 miles or more through Asia, across the Bering region and then across North America, or by an imaginary land bridge spanning the Atlantic. But if there were no North Atlantic ocean in Carboniferous times, these localities would be only about 2,000 miles apart, with solid land between them; a very possible situation for closely related faunas. A modern parallel might be the faunas of land-living vertebrates now inhabiting various parts of North America or those of eastern and western Europe.

The same considerations hold when we advance several million years up the geologic time scale. In the Lower Permian red beds of Texas are found large labyrinthodont amphibians and early reptiles, and some of these same forms are repeated in the Lower Permian Autun beds of France. The strange pelycosaur reptile *Edaphosaurus*, characterized by elongated vertebral spines occurs in Texas and in

France. If one imagines a single continent, with no intervening North Atlantic, the presence of *Edaphosaurus* and associated vertebrates can be explained, for they were members of closely related faunas, perhaps 3,000 miles apart within a single continental region.

Again, in Lower Triassic sediments, more than 200 million years old, similar fossils are found in the Buntsandstein rocks of southern Germany and the Moenkopi Formation of Arizona. In both of these regions there occur the prolific trackways of a reptile, the footprints having been given the name of *Chirotherium,* together with the bones of labyrinthodont amphibians. And in the Upper Triassic the sediments of southern Germany and North America contain numerous bones of gigantic labyrinthodont amphibians of the genus *Metoposaurus,* and

FIGURE 34. A portion of Laurasia, and some Permian tetrapods. The map shows fits between the continental blocks, with the continental shelves indicated by solid black, overlaps by dots, and gaps by horizontal lines. As is evident, there was probably a broad avenue for intercontinental migrations of land-living animals, which would explain the presence of closely related labyrinthodont amphibians (A), and the fin-backed pelycosaur reptile *Edaphosaurus* (B), in Texas and in central Europe.

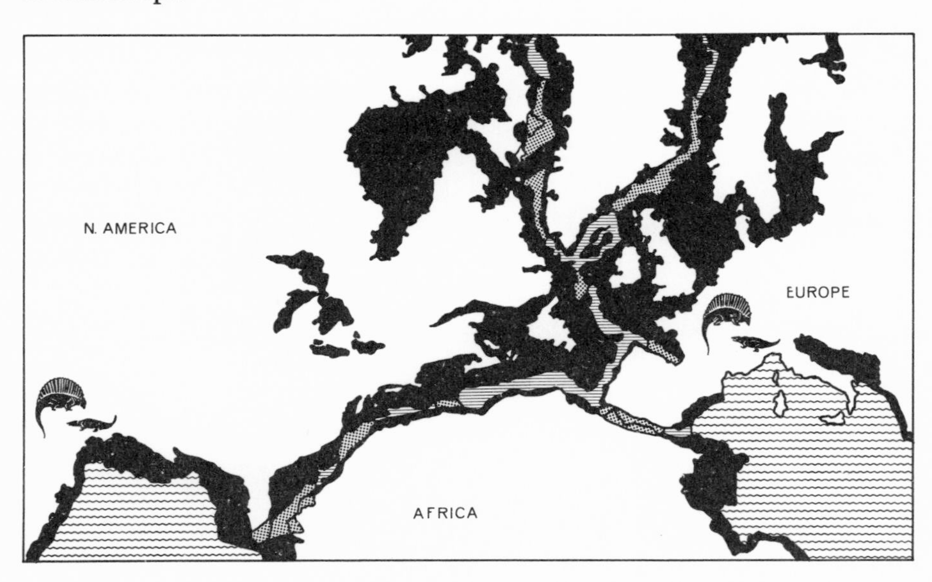

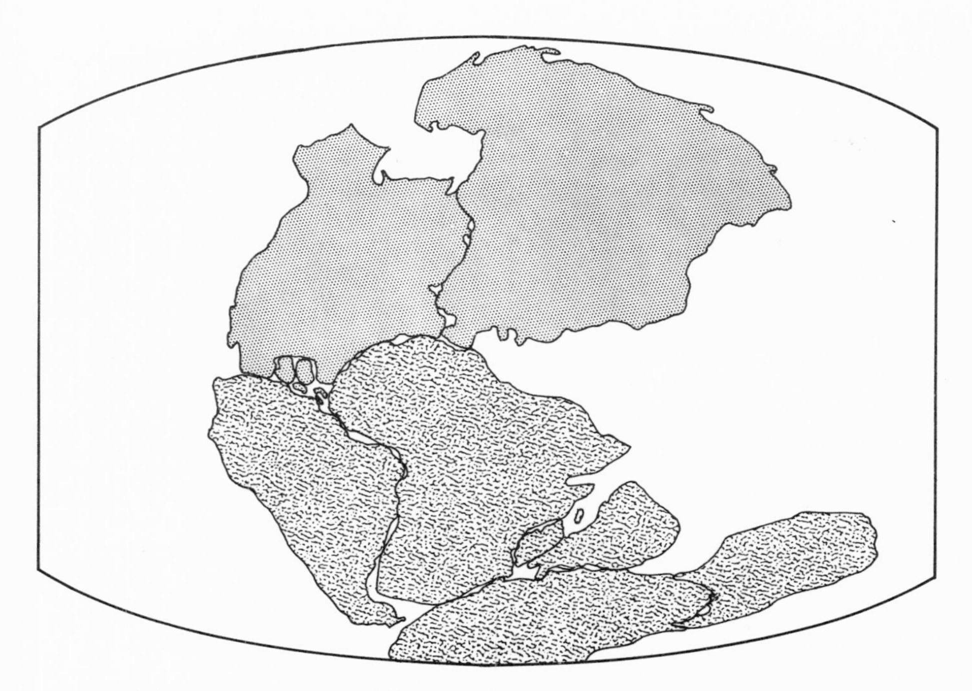

FIGURE 35. A reconstruction of Pangaea during Permian times.

large crocodilelike reptiles known as "phytosaurs" (which we will look at in more detail a little farther on), so similar in their anatomical features that the differences between them can only be on the specific level. Again, a ligation between Europe and North America is indicated.

And so there has been an accumulation and an integration of evidence during the past decades to show that there probably was a northern supercontinent, Laurasia, to complement the southern land mass, Gondwanaland. If, however, these two great ancient continents did exist, and all of the facts would seem to indicate that they did, what were their relationships to each other?

THE ONE WORLD OF PANGAEA

It is generally considered that the two supercontinents were partially isolated one from the other by a long, narrow east-to-west seaway, which has been named "Tethys." A connection obviously existed between the two supercontinents, and some believe the connection was of such nature that the two land masses formed essentially one immense supercontinent, which has been called "Pangaea."

The concept of Pangaea was proposed by Wegener a decade or so after he began his studies on continental drift. And since that time the idea of Pangaea has been variously received by the geologists and their fellow students who have concerned themselves with ancient continents and continental drift. Whether there was or was not a continent of Pangaea depends to a large extent upon what may be accepted as the limits of the Tethys Sea.

Eduard Suess proposed the name "Tethys" to designate a great seaway, which was envisaged as encircling the middle portion of the earth in an east-to-west trend, at about the latitude of the present Mediterranean Sea, thus bordering the northern edge of Gondwanaland and the southern edge of Laurasia. There is a wealth of evidence to support a former Tethys. Certainly there was a very long east-to-west geosynclinal trough in this part of the world, and it was of long duration. The uplifted and folded sediments of southern Europe and of the Himalayas, containing abundant marine fossils that are closely correlative across many degrees of longitude, provide on a grand scale the testimony as to the presence of this ancient and enduring seaway. And because of the weight of evidence indicating the wide lateral extent of the Tethyan sea, many students have felt that it formed a formidable marine barrier between the Laurasian and Gondwanaland continents. But in spite of the thickness of the sediments that constitute the Tethyan geosyncline, and in spite of the abundance of marine fossils contained within these sediments, there is nonetheless good reason to think that either the Tethys Sea was not so extended in its lateral dimensions as some have imagined, or that it was at least bridged during portions of its existence. According to one modern reconstruction, the Tethys Sea was in effect a long embayment between the northeastern border of Gondwanaland and the eastern portion of Laurasia, an embayment closed at its western extremity by the abutment of Spain against northern Africa, as has been described. If China was a part of Gondwanaland, perhaps the Tethys Sea either was narrow and restricted or was completely absent at the easternmost limit of Laurasia. In brief, Gondwanaland and Laurasia may have been joined at their eastern extremities. Perhaps one of these conditions held during late Paleozoic history. Then, as a result of the rifting of ancient land masses and the drifting of continents, there was a separation of Africa from southern Europe, so that during Cretaceous time the Tethys Sea did finally become an open ocean, stretching from the Caribbean region, along the twentieth north parallel to the eastern limits of Asia. There is a fossil record to support this view.

The Lower Beaufort rocks of South Africa, of late Permian age, contain the fossil remains of varied amphibians and reptiles, including the skeletons of large, one might almost say gigantic, primitive reptiles known as "pareiasaurs." These pareiasaurs were clumsy, bulky animals, as heavy as oxen, that evidently wandered across the terrain in considerable numbers, feeding upon plants. The jaws of their heavy skulls were provided with unbroken rows of spatulate teeth, well suited to cutting vegetation. Along with these pareiasaurs there were various carnivorous mammallike reptiles, among which were numerous genera of a group known as "gorgonopsians." The gorgonopsians were varied; some were small, no larger than house cats, some were as large as lions. They were all obviously active predators, and the largest of the gorgonopsians were something like the saber-toothed cats of a much later geological age in that they had tremendous, daggerlike canine teeth, adapted for stabbing and killing. The whole structure of the skull (as well as the postcranial skeleton) in these giant gorgonopsians was adjusted to the function of these saber teeth. It seems likely that these creatures preyed upon the clumsy pareiasaurs. This pattern of hunted and hunter is well represented in South Africa.

In sediments of the same age in the district of Perm in northern Russia (the region from which the Permian period takes its name) there are also pareiasaurs and giant gorgonopsians. These animals belong to genera different from those found in South Africa. Nevertheless the animals in Russia and in South Africa are closely related and thus give clear evidence that there must have been a terrestrial route of migration between these regions, now some 8,000 miles or more distant from each other. In fact, the two areas probably were just about as far apart during Permian time, and this may account for the generic differentiation between the animals of northern Russia and those of southern Africa. The important fact is that there was obviously a line of overland communication between the two regions during the later stages of Permian history, and this would favor the idea of a Pangaean continent, with the connection across the Iberian-Moroccan region (as well as farther to the west). Within this context it might be said in passing that discoveries made in recent years show a close relationship between the latest known Permian reptiles from North America and those of Russia, thus indicating the east-to-west unity of the Laurasian part of Pangaea.

In describing the distribution of *Lystrosaurus* during early Triassic time, it was mentioned that this wide-ranging reptile is found not only in South Africa and Antarctica, but in India as well, and on into

FIGURE 36. There were obvious north and south connections between Laurasia and Gondwanaland during late Permian history. Large, herbivorous reptiles known as pareiasaurs and their carnivorous reptilian antagonists, the gorgonopsians, inhabited what are now north-central Russia and South Africa. The very close relationships of these reptiles, the hunted and the hunter, indicate lines of access between the two regions.

China. If China was a part of Laurasia, it would seem that there were long migrations of *Lystrosaurus* and other reptiles from Gondwanaland into the northern continent. But if China belonged to Gondwanaland, as has been tentatively proposed, then the distribution of the *Lystrosaurus* fauna constitutes a continuous and eminently logical range of this early Triassic assemblage.

More might be said about Gondwanaland, Laurasia, and Pangaea, but perhaps enough has been set down to indicate the nature of the problem and some of the modern evidence pointing to the reality of these concepts of ancient supercontinents.

In summary, it does appear that what are now the several modern continents were connected in late Paleozoic and early Mesozoic time to form two great land masses, one in the southern and one in the northern hemisphere. Moreover, it does appear that these two land masses were very probably intimately connected at that stage of earth history, to form in effect one immense supercontinent, Pangaea. There is much evidence, geological and paleontological, to indicate this. The fossils of land-living vertebrates show that in those distant days amphibians and reptiles inhabited one world—a world of such dimensions that the preponderant number of animals never extended their ranges across its extent. But some of them did wander widely, and it is to them that we are indebted for this evidence of the one world of Permo-Triassic times. Let us now look at the inhabitants of this one world.

III. ANCIENT CONTINENTS AND TETRAPODS

ANCIENT LANDS AND ANIMALS

Up to this point the reconstructions of an ancient Pangaea, of Laurasia and Gondwanaland, have been presented largely upon the basis of early Triassic evidence. This temporal episode in earth history, occurring between about 225 and 195 million years ago, has been chosen because the fossil tetrapods so far discovered in Antarctica are of early Triassic age, and, as has been pointed out, these fossils from the south polar continent are of unusual significance as they bear upon the problem of the relationships of continents at the beginning of Mesozoic history. If there have been doubts as to the reality of an ancient Gondwanaland, it would seem that these doubts should be dissolved beyond much question, by the discovery within a few hundred miles of the South Pole of Triassic amphibians and reptiles, repeating in remarkable detail the contemporaneous amphibians and reptiles in southern Africa. These south polar Triassic tetrapods—the tetrapods, it will be recalled, being the four-footed animals with backbones, which in Triassic time were amphibians and reptiles—are the hard fossil evidence that Antarctica and Africa probably were parts of a single land mass; in other words, they were portions of an ancient Gondwanaland.

Gondwanaland thus may be confirmed, at least to the satisfaction of many, but this confirmation has so far been concerned with space—a reconstruction of that ancient continent as it may have existed at the beginning of Mesozoic history. There is also a time dimension to be considered; the history of Gondwanaland was long, as was the history of its counterpart, Laurasia. Let us therefore look at certain aspects in the development of the ancient continents, with particular attention given to their backboned, land-living inhabitants—what these animals were, how and where they lived together, and what the world in which they lived was like. To do this we must journey back in time, far beyond the lower limits of the Triassic period, even beyond the years of late Paleozoic times, to the transition from Devonian to Carboniferous history.

THE FIRST AMPHIBIANS

The beginnings of tetrapod evolution were marked by the appearance of the first amphibians—descendants of specialized, air-breathing fishes. These first amphibians—rather massively built animals, two or three feet in length, with heavy solid skulls, short, stout limbs, and tails that retained the fin rays that are so typical of the fishes, still closely tied to the rivers and lakes in which their immediate ancestors had lived, and still showing in many aspects of their anatomy the imprint of their direct descent from fishes—probably were widely dis-

FIGURE 37. The beginnings of the tetrapods, as exemplified by the Upper Devonian amphibian *Ichthyostega*, from East Greenland. Although this was a well-adapted land animal, as we can see by its powerful limbs, it retained various fish characters in the pattern of the skull bones and in the deep tail with fin rays. (Quite recently indications of ichthyostegid amphibians have been found in Upper Devonian sediments in Australia, an indication of the unity of Pangaea at that time.)

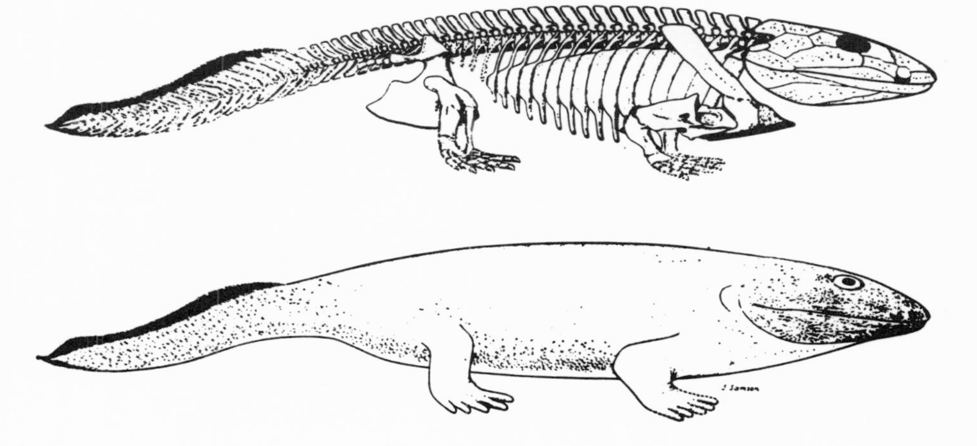

tributed across the face of the earth. However, their remains have to date been found only in high northern latitudes—on the southern shore of the Gaspé Peninsula in Quebec and particularly in eastern Greenland.

We can be sure that these first amphibians, known as "ichthyostegids" or "ichthyostegalians" (names indicative of the fishlike topography of the roofing bones in the skull), did not live in polar or subpolar climes, but rather were tropical or subtropical animals. Their occurrences today in the eastern portion of Canada and in Greenland evidently are indications of the former close connections of these two regions, as shown by the continental-shelf outlines in the North Atlantic region, as well as by various other lines of evidence that have been brought forward from recent continental-drift studies. And in those distant days, at the close of Devonian history, some 350 million years ago, when Laurasia, as an integral part of Pangaea, occupied a portion of the northern hemisphere, northeastern Canada and Greenland very probably were at considerably lower latitudes than is the case today. In fact, if the reconstruction of the continents as they may have existed prior to the formation of the Atlantic Ocean is valid, the Gaspé Peninsula would have been some 500 miles from the Mauretanian coast of North Africa, while that portion of eastern Greenland in which the fossilized remains of ichthyostegids are found, would have been perhaps 2,000 miles from this same African coast. It was a different world and, to a large extent, a tropical and subtropical world.

ANIMALS OF THE ANCIENT COAL SWAMPS

This world of benign climates and primitive land-living vertebrates persisted into the long Carboniferous period of earth history (the time span that commonly is divided by American geologists into two periods, the Mississippian and the Pennsylvanian), when great coal forests covered vast areas within the predrift continents. And it is in rocks of Carboniferous age, having a time span of about 65 million years, the beginning of which is separated from us by perhaps considerably more than 300 million years, that we find the first extensive fossil remains of land-living vertebrates. These fossils frequently occur in associations containing considerable varieties of early tetrapods—associations that with some justification may be called "fossil faunas."*

* It should be emphasized here that associations or assemblages of fossils found together in close contiguity are almost never faunas in the true sense of the word. The processes of burial and fossilization are such that there has been, in most cases, a considerable amount of selection involved in the preservation of these records of ancient animal life. Conse-

Consequently it is possible to study the distributions of Carboniferous tetrapods in a limited way, yet with some degree of assurance. This is in contrast to the occurrences of the late Devonian amphibians, the ichthyostegids, which at best are at rather isolated discovery sites, giving no comprehensive view of the distributions of the first land-living vertebrates throughout the Devonian world.

Among the Carboniferous tetrapod faunas to be preserved in the fossil record are those found in North America at Joggins, Nova Scotia, and at Mazon Creek, Illinois, and Linton, Ohio, and in Europe in the Coal Measures of Ireland and Britain, and at Nýřany and Kounová in Czechoslovakia. The fossils at these localities are found in black shales, and, as might be expected, they are overwhelmingly composed of the remains of labyrinthodont and other amphibians. During this early stage in the evolution of the vertebrates the amphibians generally were dominant, and the reptiles, such as were present, usually were small, primitive forms. There was probably some bias in the preservation of these faunas, because they seem to represent swamp-dwelling animals. It is possible that reptiles were more numerous in those days than would seem to be indicated by these Carboniferous tetrapod faunas; perhaps the reptiles were living in considerable numbers on the higher "uplands" above the coal-forming swamps and thus are not adequately represented in the fossil record. It is, however, quite probable that much of the land was low in Carboniferous times, not far above sea level, with swamps predominant on these predrift continents. Such habitats would have favored amphibians as against reptiles. So it would seem likely that during Carboniferous times the amphibians held sway for a comparatively short interval (as contrasted with the full sweep of the fossil record) before the reptiles had become well established as the varied rulers of ancient continents.

These faunas, among the earliest good representations of ancient

quently the fossils that may be found together at any particular locality, in a single geological horizon, are usually assemblages that represent only a part of the complete fauna inhabiting the region when these petrified bones were parts of living animals. For the sake of convenience, however, it has been common practice to refer to such collections as faunas, with a realization, of course, that they almost certainly represent incomplete faunas; that there are missing the remains of various animals which, when present, rounded out these aggregations so that they were balanced within the ecology of their times. To avoid clumsy circumlocutions the common practice will here be followed, and fossils that are found together in reasonable abundance and variety in one deposit or in one general region will be referred to as "faunas." Such "faunas" legitimately may be compared with each other, because quite commonly the factors that operated in one region to preserve certain animals and not others similarly operated in other regions as well.

assemblages of amphibians and reptiles in the fossil record, are in the upper portions of the Carboniferous sequence. The lower portions of the Carboniferous, those sediments included by American geologists within the Mississippian period of earth history, are almost devoid of the remains of land-living vertebrates. Early Carboniferous fresh-water fishes are known from various localities, and there are some nicely preserved Mississippian footprints, but the bones of terrestrial vertebrates are rare. In Scotland occasional amphibians of this age are found.

If the foregoing remarks seem less than satisfactory to the reader, some indulgences must be asked for. It seems to be the fate of the geologist and the paleontologist to be surrounding their statements with all sorts of qualifications. It looks like hedging. But one must realize that there are many imperfections in the paleontological record, particularly in those portions that deal with the early histories of various groups of organisms. Thus the record of land-living vertebrates during their early years is more sparsely documented than the later phases of this evolutionary story; the closer we get to the present time, the more abundant is our fossil evidence. In dealing with these early tetrapod faunas we are looking at a record that is all too incomplete.

So with such qualifications in mind, let us get back to the problem of comparing the amphibians and reptiles of Carboniferous time. Even though there are certain differences in age between the middle and late phases of the Upper Carboniferous, there are still striking resemblances between the coal-swamp faunas of North America and Europe. In making comparisons of these faunas it may be helpful, or at least interesting, to look at the manner of their occurrences.

Among the oldest of these faunas is the one found near Joggins, Nova Scotia, where the cold waters of the Bay of Fundy, pulsing back and forth in immense tidal washes—the greatest in the world—sweep against the western shore of the province. The margin of the bay is here formed of Carboniferous sediments that rise as enclosing cliffs and that slope out beneath the water to form the sea floor. The occurrences of the amphibians and early reptiles that constitute the Joggins fauna are unusual and interesting, for these fossils are found for the most part within fossilized tree stumps, some of which are contained in the steep cliffs back of the high-tide line, some of which are enclosed in the sediments along the shore, to be exposed only when the tide goes out, and some of which are uncovered in the roofs of coal-mine tunnels that angle down along the dipping sediments beneath the floor of the bay. The fossilized stumps are of lycopsids, the primitive scale trees that were so characteristic of the ancient coal swamps and commonly are preserved *upright* in the black shaly sediments. It would appear that in the Carboniferous forests of this region numerous trees would die from natural causes, as trees always have done in forests, that eventually the trunks would break away from the stumps and fall over, that the stumps would remain standing and would be kept in the standing position by the accumulation of mud around them, and that in time the interiors of these stumps would decay and would thus become hollow. These hollow stumps formed perfect traps for the small animals that lived in the ancient coal forests of Joggins; amphibians and reptiles occasionally would fall into the hollow stumps, and sometimes would be unable to climb out. There they would be buried in their tubular prisons by muddy sediments that were washed around and into the hollow stumps, and there they would be fossilized.

The Linton and Mazon Creek faunas, somewhat later in age than the assemblage at Joggins, are found in circumstances quite different from those existing along the shores of the Bay of Fundy. The fossils from Linton come from an old mine in Ohio, near the western bank of the Ohio River, about 40 miles to the northwest of Pittsburgh, Pennsylvania. The collecting of Linton fossils is prosaic indeed, as compared with the exciting search among the ancient tree stumps of Joggins; the Ohio specimens have been gathered together in large part by a patient search of the old mine dumps. Nevertheless, the fossils from Linton are beautifully preserved and have yielded rich information about vertebrate life in the Carboniferous swamps and ponds of this area.

At Mazon Creek, Illinois, the fossils are found in ironstone nodules that weather out of the Carboniferous shales along the bed of the creek. A large proportion of the fossil-bearing nodules, when broken open, reveal plant remains, a considerable number contain insects and other arthropod remains, some contain fragments of fishes, a few yield molluscs, and very occasionally a nodule will produce an amphibian. Such a comprehensive range of fossils gives a glimpse of Carboniferous swamp ecology that is extraordinarily useful in reconstructing continental life of that distant geological period.

Such a reconstruction shows us a world that was tropical, flat, and for the most part not far above sea level. The lands were extensively swampy, and within these swamps as well as on their edges there was a lush and monotonously green cover of plants: primitive trees, tree ferns, club mosses, horsetails, and ferns. There were no bright flowers to give color to the landscape or grasses to cover the higher parts of the land; many millions of years passed before these higher plants appeared. In the swamps lived fresh-water fishes and varied labyrinthodont amphibians, while small, very primitive reptiles haunted the shores and shallows at the edges of the swamps. Ancient insects flew through the air, among them giant dragonflies having wingspans equal to those of our familiar jays or crows. Other primitive insects shared the land with the amphibians and reptiles, among them cockroaches as large in their own way as were the contemporary dragonflies. It was a land full of life—but life far removed from us, not only in time but also in the nature of its expression.

Among similar Carboniferous deposits and fossils found in Europe, the Lower Coal Measures of Ireland, specifically of Kilkenny, perhaps may be equated with the Joggins deposits and their included fauna; the Middle and Upper Coal Measures of Britain, especially as they are exposed in the vicinity of Newcastle, may be generally comparable to the Linton and Mazon Creek deposits and faunas.

But perhaps the most striking of the Carboniferous tetrapod faunas in Europe are those in Czechoslovakia, specifically those of Nýřany and Kounová, which occur near the top of the Carboniferous sequence—not far below the boundary between the Carboniferous rocks and those of basal Permian age. Indeed it is evident that there was continuous deposition of sediments in this region from Carboniferous into Permian times.

The rocks, preserved in a series of small, isolated, shallow basins and consisting of carbonaceous shales and cannel coals (coals containing much volatile matter), enclosed within sandstones, shales, clays, and limestones, are very fine-grained, so that the skeletons of small amphibians and primitive reptiles are preserved in detail, thereby affording a superb record of Carboniferous vertebrate life from central Europe. It is obvious that we see in these basin deposits a record of sediments that accumulated in a low-lying continental region, where swamps and small lakes or ponds were surrounded and invaded by a luxuriant growth of vegetation. The Nýřany beds are situated to the southwest of Prague; the somewhat later Kounová beds are some 35 miles to the northwest of the city. The fossils from the latter locality were found in mine deposits, which for many years supplied coal for the gasworks of Prague.

The correlative relationships of these Carboniferous faunas of North America and Europe may be expressed in the following fashion, with the oldest ones at the bottom, the youngest at the top of the chart.

CARBONIFEROUS VERTEBRATE-BEARING ROCK OF LAURASIA			
	North America	Europe	
	Lower Wichita, Texas Lower Dunkard, West Virginia	Kounová, Nýřany,	Czechoslovakia
Carboniferous	Mazon Creek, Illinois Linton, Ohio	Upper and Middle Coal Measures,	Britain
	Joggins, Nova Scotia	Lower Coal Measures,	Ireland

THE WIDE DISTRIBUTIONS OF LAURASIAN TETRAPODS

The common denominator among the Carboniferous deposits on both sides of the North Atlantic ocean is the general similarity of the sediments, which frequently are fine-grained, black, carbonaceous shales and coals, obviously accumulated in low-lying swamps and ponds. The common denominators among the faunas are the preponderance of small amphibians, among which labyrinthodonts are prominent, the frequent goodly complement of fresh-water fishes, and the sparse presence of primitive reptiles.

The resemblances of the animals making up these several faunas have been noted by various students over the years. Details might be listed at some length, but here will be avoided. Suffice it to say that the relationships among fishes, amphibians, and reptiles in the Carboniferous sediments of Europe and North America embrace various larger categories of these animals and in numerous instances extend down to the level of genera. What does this mean?

For many years the close resemblances of the late Carboniferous amphibians and reptiles of central Europe and those from the late Carboniferous or very early Permian of North America, both zoologically and ecologically, have caused a considerable amount of argument among paleontologists of various persuasions. Texas and Bohemia are about 6,000 miles apart, in a great circle route across the North Atlantic. In the other direction, by way of northern Asia and a Bering crossing, the two regions are at least 10,000 miles apart. How, then, are the similarities of the late Carboniferous tetrapods in these regions to be explained? A connection across the North Atlantic would entail a tremendously long land bridge—a bridge that in geological terms would seem to be beyond the realm of the possible. But a relationship established by extended overland migrations between Europe and North America through northern Asia and across a Bering bridge would seem to be farfetched also, particularly in view of the close similarities of the vertebrates in the two widely separated regions. Could the animals that constituted these ancient faunas of Europe and North America, animals that were conspicuously small and relatively sedentary, have made a long intercontinental migration, equal to about half the circumference of the earth, and still have maintained the close similarities that are evident between them?

If, however, one can visualize a Laurasian continent during late Carboniferous times, in which Greenland was in close contact with North America and in which the continental shelf of northern Europe was contiguous to the eastern shelf of Greenland, then the difficulties of explaining the faunal similarities that are so evident between the late Carboniferous terrestrial vertebrates of central Europe, on the one hand, and of North America, on the other, largely disappear. All of these ancient vertebrate faunas would have lived within a single continental area. The separation between them would have been on the order of about 3,000 miles, which is considerable, but not excessive. The intervening distance would have been across a low land, covered with coal forests and interspersed with numerous swamps, and this would have allowed movement back and forth for these land-living tetrapods. Moreover, such a configuration of late Carboniferous land masses would explain the similarities in the ecology of the two areas in which these Carboniferous tetrapods were living. They inhabited essentially similar regions in the wide reaches of a broad continent. All of the evidence would seem to indicate that environments were much more uniform throughout the world in those days than they are today; consequently a separation of 3,000 miles between the tetrapods of central Europe and central North America would have been of relatively small zoological or ecological significance. It was one world in this northern sector of the globe, and the amphibians and reptiles of that distant past were part of that world.

One will note on the chart on page 93 the names Wichita and Dunkard, which up until now have not been involved in the comparison of North American and European Carboniferous faunas. These two faunas are today generally considered as of earliest Permian age, but this assignment is, in part at least, perhaps a matter of convenience rather than a definite fact. Within recent years various authorities have regarded the Wichita beds, which are found in northern Texas, and probably the Dunkard sediments as well, which occur in the tri-state region of West Virginia, Ohio, and Pennsylvania, as straddling the line of demarcation between the uppermost part of the Carboniferous and the lowermost portion of the Permian periods. This is in essence an indication of the unending flow of time and of the evolution of life. Geologic periods and other units are man-made devices for subdividing the continuum of time and evolution and as such should hardly be regarded as sacred.

The reason for the introduction of the Wichita and the correlative Dunkard beds into the final stage of Carboniferous history here is that there would seem to be rather close and significant similarities between the faunas of the Wichita sediments and the Kounová beds.

For example, at Kounová there have been found various freshwater fishes, including lungfishes that were able to burrow into the mud of lake bottoms, thus encysting themselves so that they might survive periods of drought. Also there were amphibians and pelycosaurian reptiles, including the genus *Edaphosaurus*. This is much the same association of ancient vertebrates typical of the Wichita beds of Texas. *Edaphosaurus*, briefly mentioned in the preceding chaper, is especially interesting. This was a rather large reptile, eight or ten feet in length, with an elongated body supported by strong limbs. The tail was long and strong. The head was rather small in comparison with the size of the body; the jaws were "weak" and armed with numerous small teeth. There were also large plates in the palate, covered with teeth. Obviously *Edaphosaurus* was not an aggressive predator, as were so many of its pelycosaurian relatives, but rather was a harmless reptile, feeding possibly upon plants or possibly upon fresh-water molluscs. But the most striking feature of *Edaphosaurus* is the tremendous elongation of the spines of the vertebrae, which in life evidently supported a high, membranous "sail" down the middle of the back. Not only were the vertebral spines elongated, but also they were decorated with somewhat irregularly arranged crossbars, like the crossbars of telephone poles. Just what might have been the purpose of such bizarre decorations is a question presently beyond our ken. It does seem possible that the sail of this reptile, as was the case in other pelycosaurs having such sails, served as a sort of temperature-control device—to absorb heat from the sun when the animal was cold and to radiate heat when the animal was too warm. The presence of this remarkable reptile in central Europe and in Texas makes a striking demonstration of the ready avenues for migration between the two regions during the final days of Carboniferous time and the beginning of Permian history.

The important point in this connection is that the essential com-

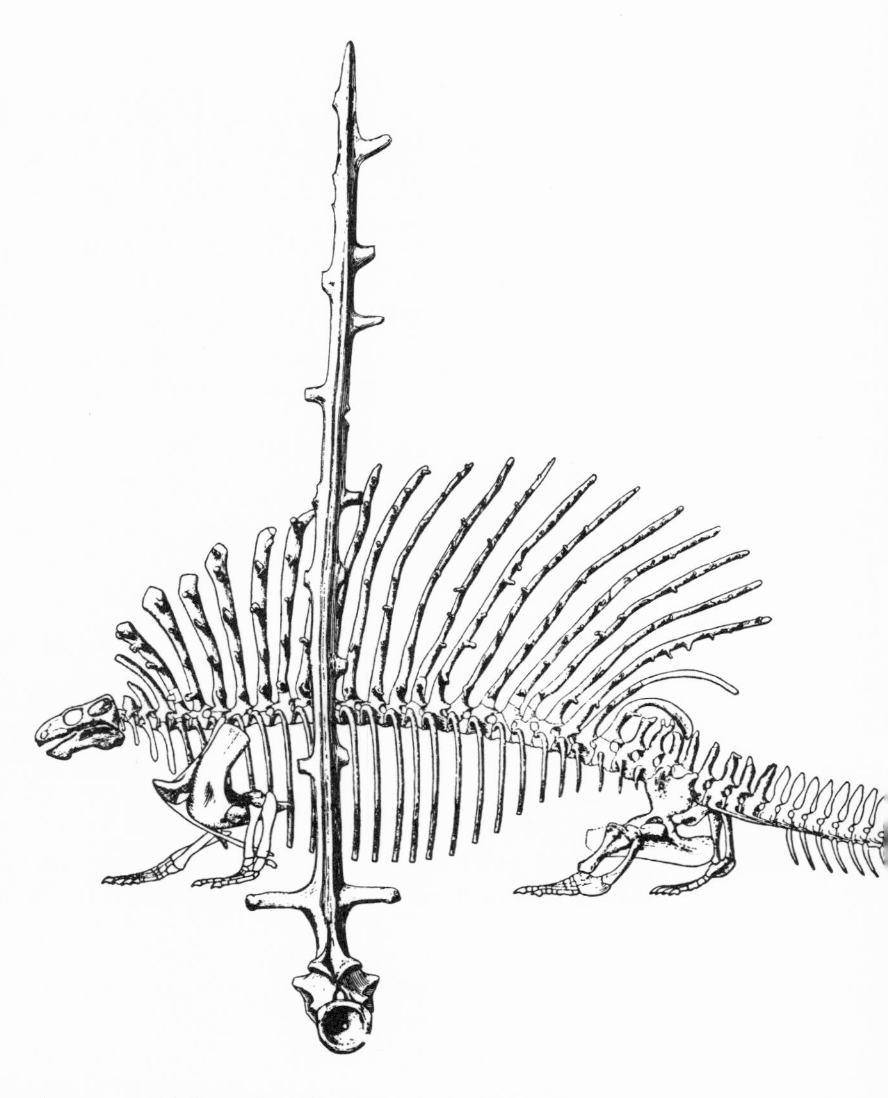

munity of tetrapod life in North America and Europe extended to the very end of Carboniferous time, and beyond. Which brings us to the Permian period.

PERMIAN ROCKS AND FOSSILS

The occurrences of land-living vertebrates in Permian rocks, ranging in age from about 280 to 225 million years ago, are much more numerous and widespread than is the case of the earlier Carboniferous fossils. Consequently our knowledge of tetrapods in relationship to

FIGURE 38. A close look at the bizarre reptile *Edaphosaurus*, an inhabitant of North America and Europe during early Permian times. The presence of such a uniquely specialized reptile in Texas and France indicates a broad corridor for movement of land-living animals between the two regions.

Permian continents is extended to a degree that is not possible in the consideration of Carboniferous amphibians and reptiles. Permian rocks containing tetrapod fossils are found widely distributed throughout the world; perhaps a chart will be helpful to introduce the names of formations and faunas discussed in the following pages.

Before entering upon a discussion of Permian tetrapods and continents, certain points of difference between the animals and the lands of Permian age and those that we have looked at of Carboniferous age should be emphasized. In the first place, the land-living vertebrates of the Permian differ from those of the Carboniferous by the preponder-

PERMIAN VERTEBRATE-BEARING ROCKS

	North America	Europe	Africa	Other Continents
Upper		Dvina, Zone IV, Russia Cutties Hillock, Britain Zechstein, Germany	Lower Beaufort, Daptocephalus zone Cistecephalus zone, S. Africa Ruhuhu, E. Africa	Bijori, India
Middle	Flower Pot, Texas San Angelo, Texas Hennessey, Oklahoma	Dvina, Zones I–III, Russia Kupferschiefer, Germany	Lower Beaufort, Tapinocephalus zone, S. Africa	
Lower	Clear Fork, Texas Abo, Cutler, N. Mexico Wichita, Texas Dunkard, Tristate region	Autun, France Rotliegende, Germany	Ecca, S. Africa Dwyka, S. Africa	Itarare, Brazil

ance of medium-sized to large animals, as compared with the generally small animals that preceded them. Moreover, there is a large proportion of reptiles among the Permian tetrapods, as contrasted with the overwhelmingly amphibian complexion of the Carboniferous faunas. Again, Permian tetrapods are very commonly found in red-colored sediments—sandstones and mudstones—in distinct contrast to the black, carbonaceous shales in which Carboniferous tetrapods are commonly preserved. These facts are worth remembering, for they reflect the trends of vertebrate evolution and of continental developments. The land-living vertebrates were "growing up" in the evolutionary sense; they were to a considerable degree becoming giants, as compared with their Carboniferous predecessors, because it was in every way advantageous for them to be giants of a sort. And the continents were growing—in the sense that they were becoming on the average more elevated than were the lands of Carboniferous times. Rolling uplands and varied topographic features were replacing the low swamps of the age of coal forests. These evolutionary developments in the lands, and among the animals that lived upon them, were to make of the Permian a period quite different in its aspects of land and life from the Carboniferous period.

PERMIAN TETRAPODS OF LAURASIA

Whether or not one considers the Wichita (and the Dunkard) beds as extending down into the Carboniferous, as already has been suggested, is a matter of personal opinion and one of no very great import. Certainly these beds to a large degree contain the beginning of Permian vertebrate history in North America. Above the Wichita in Texas are the Clear Fork beds, in which the faunal developments, so amply recorded in the Wichita, are carried on and magnified. And above the Clear Fork are the San Angelo and Flower Pot formations with contained vertebrate faunas. The Abo and Cutler formations in New Mexico reflect the faunal expressions of the Texas sediments.

These Lower Permian rocks in Texas and adjacent states are often referred to as "red beds" because of their characteristic colors, ranging from brick red to rather brilliant hues. These are sandstones and mudstones, obviously deposited in streams and ponds, and the evidence would seem to indicate that this whole general area was a low delta, fronting upon an arm of the sea, which in those distant days extended from the south to separate Texas from New Mexico, much as today the Gulf of California separates Baja California from the Mexican mainland.

The Texas red beds have long been famous for the fossil amphibians and reptiles found in them. The amphibians are dominantly labyrinthodonts, but there are also other amphibians, notably certain highly adapted dwellers of streams and lakes known as "nectridians,"

FIGURE 39. The lower Permian red beds of north-central Texas. These sediments have yielded abundant fossil amphibians and reptiles, to give us a comprehensive view of vertebrate life in Laurasia at the beginning of Permian time, as shown in figure 40.

some of these characterized by flat, very broad, arrowhead-shaped skulls and weak limbs; and some elongated amphibians known as "aistopods." The labyrinthodonts were various, and some of them were quite large; *Eryops* was a veritable amphibian giant, with a large, heavy skull and rather stout limbs. Amphibians of this type probably were active competitors with the reptiles of their time, and perhaps they represent a last expression among these vertebrates as dominant members of the faunas to which they belonged.

The red beds are noteworthy not only because of the giant amphibians found in them, but also because of the numerous large reptiles contained in these sediments. Here, for the first time in geologic history, the reptiles become predominant among the land-living animals. Although the red-bed reptiles are various, they are included largely within two orders, the cotylosaurs and the pelycosaurs. The cotylosaurs were stem reptiles, descended directly from amphibian ancestors. Some of these, known as "captorhinids," were small and rather lizardlike in their ecological adaptations. But others were large and clumsy. The rulers of the red-bed faunas, however, were the pelycosaurs, the ancestors of the mammallike reptiles. We have just looked at *Edaphosaurus,* the strange fin-backed pelycosaur that ranged from Texas into central Europe during late Carboniferous and early Permian time. There were many other pelycosaurs in the Texas red beds, however. For the most part they were active predators, often ten feet or more in length. They commonly had large skulls with long jaws, these set with sharp, bladelike teeth, all in all a mechanism for striking down and eating other large reptiles or amphibians. Some of these pelycosaurs had huge sails along the middle of the back, similar in a way to the sail of *Edaphosaurus,* but lacking crossbars on the spines that supported the sail. *Dimetrodon* was such a reptile, truly the ruler of the early Permian scene.

In the early Permian Autun beds of France as in the late Carboniferous of Kounová are found labyrinthodonts of a cast similar to those of the Texas red beds, as well as cotylosaurian and pelycosaurian reptiles, including fin-backed forms among which is present the genus *Edaphosaurus,* so characteristic of the Texas Permian. Of course, such similarities are indicative of the continuing community of faunal relationships between North America and Europe, and correlatively the continuing broad continental connections between these two regions which today are so effectively separated by the width of the North Atlantic Ocean.

FIGURE 40. A glimpse of Laurasia during early Permian time, as revealed by the red bed fossils of Texas. On the upper left is *Diadectes,* a herbivorous cotylosaurian reptile (or perhaps an amphibian). In the water are some amphibians, the large labyrinthodont *Eryops* and the small nectridian *Diplocaulus,* with a broad, flanged skull. *Ophiacodon,* the fish-eating pelycosaur, is at the edge of the pond, and to the right are the two fin-backed pelycosaurs—*Edaphosaurus,* a herbivore, and *Dimetrodon,* a large, aggressive carnivore. Between the two latter is *Cacops,* a labyrinthodont amphibian. In the lower right corner is *Captorhinus,* a small, carnivorous cotylosaur that may have fulfilled after a fashion the role of a lizard in the early Permian scene.

The junction of North America and Europe, so apparent in the similarities of the coal-swamp faunas of the Carboniferous, and the red-bed faunas of the Lower Permian, in which the large pelycosaurian reptiles were so prominent, evidently was continued into the middle portions of Permian time. The Hennessey, San Angelo, and Flower Pot faunas, the last of the Permian vertebrate faunas in North America, although rather fragmentary, show that the connection with the European region still persisted. These faunas are noteworthy because of the large sizes among the reptiles composing them; such reptiles as cotylosaurs and pelycosaurs and perhaps a mammallike therapsid, displaying striking parallels to the large reptiles found in the Permian of northern Russia.

GLOSSOPTERIS AND *MESOSAURUS* AGAIN

So far our attention has been directed to the northern portion of the globe, to Laurasia, during the early history of land-living vertebrates. But what about the southern hemisphere, the site of the great Gondwana continent?

Here, unfortunately, there is no definite vertebrate evidence for the Carboniferous period, even though land areas obviously were extensive at that time. As for the Permian, there were extensive and abundant amphibian and reptilian faunas in Africa, particularly during late Permian time, but a surprising and singular paucity of evidence in other southern continental areas. Yet this was the period of the distinctive *Glossopteris* flora, the plant association which spread across Africa and South America, Australia and Antarctica, and into peninsular India, to afford such strong paleobotanical support for a continent of Gondwanaland.

Much of the literature on Gondwanaland devotes extended attention to *Glossopteris* as proof of the former connection of the several southern continents to form the great supercontinent that occupied this part of the world. And much of the literature is also concerned with *Mesosaurus*, the Permian reptile discussed in the preceding chapter, found only in South Africa and in Brazil.

In South Africa, in the Great Karroo Basin, which forms the semi-desert and grasslands between the folded mountains of the Cape and Johannesburg, the Permian sequence is composed of the Dwyka beds at the base, showing clear evidence through most of their thickness of glaciation, the unfossiliferous Ecca beds above them, and on top of the Ecca, the Lower Beaufort group, with abundant and marvelous fossils of amphibians and reptiles in great variety. The top portion of the Dwyka consists of limy shales, at the top of which is the prominent White Band, as it is called, running like a great ribbon, mile after mile, along the undulating folded exposures of this geological formation. The Dwyka shales contain *Mesosaurus.*

Mesosaurus, almost identical with its counterpart in South Africa, is found in the Itarare Formation, as it is exposed in the Brazilian states of Paraná, Santa Catarina, and Rio Grande do Sul.

By themselves, the distributions of these fossils might be questioned. In concert with all of the other evidence that has accumulated within these last few years, *Glossopteris* and *Mesosaurus* do indeed point with clarity to the reality of Gondwanaland and of continental drift.

THE PERMIAN REPTILES OF SOUTH AFRICA

The Lower Beaufort beds of South Africa, which form a magnificent, thick sequence of mudstones and sandstones above the Dwyka Formation, with its White Band and *Mesosaurus,* and the paleontologically barren Ecca beds, are of middle and late Permian age. These sediments, so famous for their fossils, have been divided into three zones, named after key reptiles that are especially characteristic of each of the zones. The lowest is the *Tapinocephalus* zone of middle Permian age, and above it are successively the very fossiliferous *Cistecephalus* and *Daptocephalus* zones, of late Permian age. These characteristic genera are all mammallike reptiles, or therapsids, as indeed are the overwhelming majority of reptiles in the Beaufort sequence. *Tapinocephalus* was a big reptile, as heavy as a large ox, remarkably bulky, with a barrellike body, thick legs, and a massive skull composed of thick bones. The teeth in this strange reptile were small, and it seems evident that *Tapinocephalus* was a slow, harmless herbivore. The zone in which it is found is especially characterized by the large, clumsy therapsids contained within it. *Daptocephalus* is a large dicynodont; the zone fossil *Cistecephalus* is a small dicynodont, somewhat similar to *Lystrosaurus,* the Triassic dicynodont with which we have already become acquainted, and of the ilk that abound in these upper zones of the Lower Beaufort beds. The extensive collections of Beaufort fossils that have been made through the years show that about 85 percent of the reptiles in these sediments are dicynodonts, and a major proportion of the remaining forms are mammallike theriodonts—active, slender-limbed, predatory carnivores, characterized by rather doglike heads, in which there were sharp incisor teeth for nipping, large canine teeth for stabbing, and frequently bladelike or cusped cheek teeth for biting and chewing. Among the Upper Permian theriodonts of South Africa the gorgonopsians are numerous and varied. These were rather specialized predators, in which the canine teeth and frequently the incisors as well were commonly greatly enlarged, while the cheek teeth, behind the canines, usually were remarkably reduced in size and in numbers. In fact, some of the gorgonopsians had only four such cheek teeth, all of them reduced to mere pegs.

One may picture southern Africa during Lower Beaufort time—that is to say, during the later stages of Permian history, perhaps about 250 million years ago—as a rolling upland, with small streams and lakes between the low, rounded hills. It was a land of primitive vegetation, with many ferns covering the ground. And in this land the strange reptiles of that day found refuge and made their separate and varied livings. The slow-moving pareiasaurs and the other equally ponderous reptiles, such as *Tapinocephalus,* browsed upon the abundant plants with which they were surrounded, as did the host of smaller two-tusked

FIGURE 41. An excursion to Gondwanaland. Members of the International Gondwana Symposium of 1970 in South Africa examining the Upper Permian Beaufort beds near New Bethesda, Cape Province. From sediments such as these came the fossils that make possible a reconstruction of late Permian life in Gondwanaland, as seen in figure 42.

dicynodont reptiles. All of these plant-eating reptiles made up the broad base of the vertebrate food pyramid, as is evident from the great numbers of their fossils in the ground. And preying upon them were the different reptiles occupying the upper part of the food pyramid—reptiles such as the saber-toothed gorgonopsians and other theriodonts generally related to them. It was a land full of life, as is Africa today—but life of a different sort and on a different plane of activity. Whereas the lion today charges an antelope or a zebra seemingly with the speed of an express train, the gorgonopsian probably attacked a pareiasaur among the hills of ancient Africa at a rate that in comparison with the movements of the lion would seem to be almost leisurely. Nonetheless the actions between Permian hunter and hunted were as effective as those involving their mammalian counterparts so many million years later. The drama of life and death was there.

SOUTH AFRICA AND RUSSIA

In the district of Perm in northern Russia, the region where the pioneer English geologist Sir Roderick Impey Murchison studied the Permian system of rocks and thus established the Permian period, are found sediments correlative with the Lower Beaufort beds of South Africa. Zones I to III of the Russian sequence (see chart on page 98) may be equated with the *Tapinocephalus* zone and Zone IV may be correlated with the *Cistecephalus* and *Daptocephalus* zones.

In Zone IV of Russia is found a large pareiasaur, very closely related to *Pareiasaurus* of South Africa, and the gigantic gorgonopsian *Inostrancevia,* paralleled by the gigantic South African gorgonopsian *Rubidgea.* We have noted in the last chapter how these gigantic carnivores may have preyed on the pareiasaurs, so abundantly present in the late Permian landscapes of Russia, as they were in South Africa. We have also noted how the presence of closely related reptiles in northern Russia and in South Africa indicate an undoubted connection between these two regions, i.e., Laurasia and Gondwanaland. It had to be so; to imagine the pareiasaurs and gorgonopsians of Africa and Russia as having evolved independently goes beyond the bounds of credibility. So here is definite evidence that the two great Permian supercontinents were closely connected in late Permian time, as components of Pangaea.

The relationships of the Permian tetrapods of Russia were not, however, restricted to communications between the northern and southern land masses of that geologic period—that is, between Gond-

FIGURE 42. A late Permian Gondwana scene in what is now South Africa. It is a scene in which the mammallike reptiles, or therapsids, are dominant; the only nontherapsid being the defunct cotylosaur *Pareiasaurus* and the three pareiasaurs retreating into the distance. Straddling his pareiasaurian prey is the huge therapsid carnivore *Rubidgea*. Other therapsids confront *Rubidgea;* two carnivores known as *Scymnognathus* on the left, and a smaller one, *Lycaenops,* in the lower right corner. In the middle foreground is a small, probably insectivorous therapsid *Galechirus*. In the middle distance are two herbivorous therapsids known as dicynodonts (belonging to the genus *Dicynodon*), while behind them are some large labyrinthodont amphibians.

wanaland and Laurasia. For there would seem to have been a certain amount of movement back and forth in an east-to-west direction, between the eastern and western portions of Laurasia, which were presumably still connected regions within the great supercontinent that then occupied the northern hemisphere.

To appreciate this it is necessary to go back somewhat in time, from the late Permian, when the sediments of Zone IV in Russia (see chart on page 98) were being deposited, to what may be considered as the middle Permian, when the sediments of Zones I–III were being laid down. At about this time the fresh-water sands and clays that make up the Hennessey, San Angelo, and Flower Pot formations, to which we have been introduced, were being formed.

RUSSIA AND NORTH AMERICA

As we have seen, various late Permian reptiles in Russia and in South Africa were of giant proportions. The forerunners of these gigantic tetrapods had appeared in Africa in the lower portions of the Lower Beaufort beds and in Russia in Zones I–III. It would seem significant that in the Hennessey, San Angelo, and Flower Pot formations there are also reptiles of very large size—an indication of a trend to giantism that was well developed among the tetrapods as the course of Permian history proceeded to its concluding stages. These large reptiles in the North American sediments represent in part a course of evolution parallel to that which took place in what is now the Old World, for the large reptiles of the uppermost North American Permian beds are certainly to some degree the direct descendants of the pelycosaurs that so dominated the Texas scene during Wichita and Clear Fork times. But it is quite evident that some of the San Angelo and Flower Pot reptiles are clearly related to certain large therapsids so beautifully and abundantly preserved in the lower zones of the Russian sequence.

Therefore, it would appear that there were some east-to-west exchanges during this phase of Permian history. Interestingly, the exchanges appear to have been highly selective, and the movement would seem to have been essentially from Russia into the North American region. In other words, certain San Angelo and Flower Pot genera show their derivation from Russian forms, but the reverse seemingly is not the case.

Perhaps this appearance of Russian-derived forms in the uppermost Permian reptile-bearing beds in North America may indicate what has been called the "filter effect." Although it is probable that at this time there still was a broad connection between the eastern and western portions of Laurasia, nevertheless there may have been ecological conditions that served to act as a filter—permitting only a limited segment of the Russian fauna to move westward. For example, the Sahara Desert today forms a most effective filter, limiting drastically the interchange of mammals between North Africa and that part of the continent to the south of the desert. Some such factor, although not necessarily a desert, might have been involved in determining the limited interchange between what are now northern Europe and North America during Permian times.

Certainly it would seem evident that the possibilities for east and west movements of tetrapods in middle Permian times, limited though they might have been, nonetheless did exist. Thus the pattern of latitudinal migrations within Laurasia continued into the middle part of Permian history, as it had existed in earlier Permian times, and even before that, in the Carboniferous.

There is no record of late Permian land-living vertebrates in North America, so nothing can be said as to the east and west movements of Laurasian reptiles at the close of Permian time—the time when there was a decided interchange of these tetrapods between Russia and Africa.

Likewise, there is little to say about Permian tetrapods in those two integral parts of Gondwanaland, India and Australia. Permian sediments are present, but land-living vertebrates are sparse, to say the least, or completely lacking. In India the Bijori beds have yielded a labyrinthodont amphibian, and with this brief statement we may end our résumé of Permian tetrapods on Permian continents.

THE PERMO-CARBONIFEROUS WORLD

What conclusions are to be drawn from this survey of late Paleozoic predrift vertebrate faunas? The evidence, as we have seen, is spotty: a few Carboniferous faunas in Europe and North America, a well-documented sequence of Lower Permian faunas in North America, and likewise well-documented Middle and Upper Permian faunas in South Africa and Russia, with enough North American documentation extending into the later, but not the latest, phase of Permian history to make some comparisons possible. Finally there is some slight evidence in India and other parts of Asia, and in Australia, even less in South

America, and none as yet in Antarctica. But even though the evidence of the Permian land-living vertebrates is unevenly distributed among the present continents, it is still sufficiently well attested to give us some picture of Paleozoic life in the predrift world.

In effect the Permo-Carboniferous lands formed one world, or, at the very least, two closely connected worlds—Gondwanaland and Laurasia; perhaps it would be justified to say that these supercontinents were merely divisions of a larger Pangaea. Within this one great land mass, or two closely connected land masses, the terrestrial vertebrates evolved as four dominant groups: the labyrinthodont amphibians, the primitive cotylosaurian reptiles, the variously adapted pelycosaurs, and the progressive therapsids—the reptiles that were truly mammallike. These tetrapods ruled the faunas of which they were the constituents.

There were communities of relationships among these primitive predrift faunas, indicating the close connections of the continents during Permo-Carboniferous times. Land-living amphibians and reptiles moved back and forth between North America and Europe, and between Eurasia and Africa; very probably there were migrations between various southern-hemisphere continents as well, but as to the distributions of Gondwana tetrapods, the evidence is as yet not all in hand.

IV. THE TRIASSIC TRANSITION

THE ADVENT OF THE TRIASSIC

It would seem evident that a world in which lands took the form of two great connected supercontinents, one essentially in the northern hemisphere and one in the southern, continued from late Paleozoic times into the beginning of Mesozoic history. This was in a sense one world, yet one world the northern and southern components of which retained to a large degree their distinctive features. Gondwanaland and Laurasia, joined along portions of their respective northern and southern edges into an immense Pangaea, appear to have been as characteristic of early Triassic history as they had been of Permo-Carboniferous time. And as in the closing stages of Paleozoic history, the two massive continents, continuing into the Triassic, would seem to have harbored to a notable degree their distinctive vertebrate faunas.

Of course, varied developments were altering land and life at this stage of earth history, more than 200 million years in the past. Even though Gondwanaland and Laurasia continued in their larger aspects during the passage from Paleozoic into Mesozoic history (i.e., from the end of the Permian period into the beginning of the Triassic period) and even though the development of land-living amphibians and reptiles in the northern and southern hemispheres continued along their sometimes separate, sometimes related paths, as they had in the past, there were changes taking place across the surfaces of the two great continents—physical and biological changes. The faces of the lands were being molded anew by geological forces, forces within the earth causing the uplift of hills and forces upon the earth causing the erosion of those hills to fill valleys and the ocean floor with their detritus. Along with these physical changes were the correlative changes among the animals and plants of the continents, so that the life of Gondwanaland and of Laurasia during Triassic time took on new aspects, to distinguish it from the life of these supercontinents during preceding Permian time. This was especially marked among the tetrapods—the amphibians and reptiles—as we shall see.

(It should be mentioned that our view of early Triassic tetrapods, particularly in Gondwanaland, is considerably more detailed, and hence more convincing, than it is for the tetrapods of Permian age. Various areas in our modern world [South America, for example] that have afforded scant or even no evidence of Permian land-living amphibians and reptiles are well supplied with the fossil remains of these animals in rocks of Triassic age. All of which lends to the Triassic tetrapods a particular interest and an especial significance for the student of earth history and of organic evolution.)

Although the continents at the beginning of Triassic time reflected to a large degree the same general relationships between them that had marked late Paleozoic history and although many of the amphibians and reptiles inhabiting those continents were the direct and only slightly modified descendants of their Permian ancestors, there were signs of many changes that were to develop during the course of some 30 million years of Triassic history, to make the end of this period very different from its beginning. The Triassic period was basically a time of transition; it was the geologic period intervening between an old and a new world. It was a time when profound changes in the interrelationships of the continental masses had their beginnings, and concomitantly it was a time when the animals and especially the vertebrate faunas living on those continents were experiencing revolutionary alterations. In the history of vertebrate evolution the Triassic period was the time when various amphibians and reptiles that had held over from Permian times, as the final expressions of long-established evolutionary lines, lived side by side with many newly arisen groups of land-living tetrapods. Even though there was a considerable extinction of land-living vertebrates during the transition from Permian to Triassic times, some of the old groups of animals that had been so successful for many millions of years persisted for a sufficiently extended time in the Triassic to witness the rise and development of new groups that were to set the large patterns of tetrapod evolution for the forthcoming hundred million years. Let us consider briefly the changes among land-living tetrapods—the Permian extinctions, the Triassic holdovers, and the Triassic newcomers—that gave to Triassic tetrapod faunas their particular characters.

PERMIAN HOLDOVERS AND TRIASSIC NEWCOMERS

Various amphibians that had been typical of Permian faunas disappeared during or even before the transition from the Permian to the Triassic. Such were the many small amphibians that haunted late Paleozoic rivers and lakes, and even large numbers of labyrinthodonts, the amphibians that for so long had shared with the early reptiles the domination of the land. Nonetheless some labyrinthodonts survived the change from the Permian to the Triassic to become prominent and almost ubiquitous elements in Triassic faunas. These were the stereospondyls, large-headed, flat-skulled amphibians highly specialized for a thoroughly aquatic life in the streams and ponds of Triassic landscapes. Some of the stereospondyl amphibians were giants of their kinds—the largest amphibians ever to have lived, massive animals that attained lengths of six or eight feet.

The seymouriamorphs—vertebrates showing so many characters transitional between amphibians and reptiles that they are variously classified either on the amphibian or on the reptilian side of the line, and prominent in some Permian faunas—also had become extinct by the close of Permian history. A few other lines of evolution also disappeared at or before the close of Permian time, such as the eunotosaurs, a restricted group of reptiles—so restricted that they need not especially concern us here—that lived in South Africa during a part of Permian history, and the mesosaurs of South Africa and Brazil, so often cited as proof for the juncture of these continents during Permian times. The pelycosaurs, which we have already looked at in some detail, also failed to survive into Triassic times, but in a sense their evolutionary line was continued by the mammallike therapsids, which ultimately were of pelycosaurian ancestry.

Certainly the therapsids, as represented by the two-tusked, herbivorous dicynodonts, among which our old friend *Lystrosaurus* was present, and by the doglike or wolflike theriodonts, active and aggressive predators of Permo-Triassic times, were prominent in Triassic faunas. In fact, some of the Gondwana faunas are almost overwhelmingly therapsid in their composition. These, together with the labyrinthodont amphibians, were the most varied, the most important, and the most numerous of the Permian holdovers into Triassic time. One might think, in viewing some of the Triassic faunas, that the stereospondyl labyrinthodont amphibians and the mammallike therapsid reptiles, being so numerous and well established, would survive for countless years into the geologic future. But this was not to be the case, as we shall see.

The few other reptilian groups that held over from Permian into Triassic times were the primitive cotylosaurs, which like the labyrinthodont amphibians were in the final throes of their long history, and in addition the protorosaurs and the eosuchians. The protorosaurs were a small and relatively unimportant group of lizardlike reptiles (using this term "lizardlike" only for the sake of analogy), reptiles that continued their history until the close of the Triassic period. The eosuchians, although not important numerically, were very significant reptiles in the evolutionary sense, for from them sprang the lizards and snakes and rhynchocephalians, the thecodonts, crocodilians, flying reptiles, and dinosaurs, all of which were to dominate the lands during middle and late Mesozoic history.

And this brings us to the newcomers among the tetrapod faunas of Triassic age. The thecodont reptiles—in which the skull generally was narrow and deep, and characterized by two large openings on each side behind the eye, the vertebral column was well-articulated and strong, the limbs and feet were commonly long and slender and well-suited for rapid running, and the body was well-protected by armor plates—have just been mentioned. These progressive reptiles, descended from eosuchian ancestors, appeared essentially with the beginning of Triassic history and became extinct at the close of Triassic times. Their sojourn on the earth was short in geological terms, but important in that these reptiles were the direct ancestors of those other newcomers in the Triassic faunas that were to be the dominant reptiles of the Mesozoic, namely, the crocodilians and the two orders of dinosaurs. These various thecodont descendants made their several entrances into early Mesozoic faunas during middle and late Triassic times.

Other newcomers among the Triassic tetrapods were, in combination with the crocodilians, those animals representatives of which have survived into the modern world—the turtles and the lizards, which need no introduction, and the rhynchocephalians, numerous in many Triassic faunas, but surviving today only as the lone tuatara of New Zealand. Referring momentarily to the amphibians, the first frogs also made their appearance in Triassic times.

Mention should be made of several groups of reptiles that appeared in the Triassic, the ichthyosaurs, the nothosaurs and plesiosaurs, and the placodonts. These were all marine reptiles and represent the first true invasion of the seas by tetrapods. As such, these denizens of the open oceans are of prime importance to the student of evolution, but they will receive scant attention in this work. Our interest is centered upon the continents and the backboned animals that inhabited the continents.

THE *LYSTROSAURUS* FAUNA IN GONDWANALAND AND ASIA

With this brief review in mind, we may look once again at the reptile that has loomed so large in this story, *Lystrosaurus,* and at its contemporaries living in Gondwanaland during the early years of Triassic time. The contemporaries deserve attention here, for they were given little if any notice in the previous remarks about the discoveries of *Lystrosaurus* on the several modern remnants of ancient Gondwanaland.

The *Lystrosaurus* fauna in Africa is an assemblage dominated by amphibians and reptiles belonging to groups that held over from Permian time. It is essentially a labyrinthodont-amphibian–therapsid-reptile fauna. *Lystrosaurus* itself is, of course, a mammallike therapsid, one of that large therapsid group known as "dicynodonts," which were extraordinarily numerous in the Upper Permian faunas of Africa. With *Lystrosaurus* is *Dicynodon,* in this case a genus that persisted from the Permian into the early Triassic. There are various small theriodont reptiles (somewhat analogous to weasels and civets) in the *Lystrosaurus* fauna, these having their origins in late Permian ancestors. Of these, one genus, *Thrinaxodon,* is especially well known from a considerable number of complete skulls and skeletons. Indeed, *Thrinaxodon* is one of the most completely documented of the very mammalianlike theriodonts, showing us, in the details of its anatomy, a small, obviously active predator, with a rather doglike skull (in miniature) supplied with pointed canine teeth, and with sharp, cutting cheek teeth, and having a strong, well-knit body mounted on slender legs adapted for comparatively fast running. When one looks at the skeleton of *Thrinaxodon* one can imagine that perhaps this animal was more mammalian than reptilian in its mode of life. It may very well have been to a degree warm-blooded, with a covering of hair. *Thrinaxodon* is one of the conspicuous members of the *Lystrosaurus* fauna. Also the *Lystrosaurus* fauna contains procolophonids, which are small cotylosaurs, again of Permian origin, while among the amphibians there is an array of labyrinthodonts directly descended from Permian ancestors. These are all "holdovers," linking the *Lystrosaurus* fauna with its Permian antecedents.

But two groups of reptiles in the *Lystrosaurus* fauna mark the new Triassic newcomers—the harbingers of new tetrapods that eventually were to inherit the earth. These two new groups are prolacertids, the ancestors of the true lizards of later times, and thecodonts, the already-mentioned forerunners of the dinosaurs and their relatives. Three genera form the record of these Triassic newcomers in the African *Lystrosaurus* fauna: two of them, *Prolacerta* and *Pricea,* both rather rare, are prolacertids, in life looking very much like rather small lizards, of which they were more or less the ancestors; the other, *Proterosuchus* (commonly referred to in the literature as *Chasmatosaurus*) is a thecodont, a moderately large reptile three or four feet in length, with a long-snouted skull armed with numerous sharp teeth, the front of which formed a curious sort of hook over the equally toothy lower jaw. *Proterosuchus,* like *Thrinaxodon,* is a prominent and important constituent of the *Lystrosaurus* fauna.

The fossils from Antarctica, as we now know them, reflect the African *Lystrosaurus* fauna to an uncanny degree. *Lystrosaurus* is, of course, the key fossil, and while upon the basis of our present evidence this reptile is not so abundant as in many African localities, it would seem to be interestingly enough abundant to the degree that characterizes its occurrence in Natal, along the southeastern African coast. Perhaps this is significant, for if Africa and Antarctica were joined in Triassic time, the Natal border of Africa would be contiguous to the Antarctic land mass.

Thrinaxodon has been found rather abundantly preserved in Antarctica—a distinct link with Africa. Also the Triassic amphibians found in Antarctica would seem to indicate occurrences similar to those in the Karroo Basin. There is, however, a surprising difference between the Antarctic and African occurrences of the *Lystrosaurus* fauna. In Antarctica the prolacertids are seemingly abundant, whereas in Africa these little precursors of the lizards are, as already mentioned, relatively rare.

The differences between the *Lystrosaurus* fauna as it is found in Africa and in Antarctica are, however, differences of degree rather than of kind. Basically there are remarkably strong faunal resemblances between the two regions—a telling argument in favor of their close connection during early Triassic history. Moreover, the fact that the *Lystrosaurus* fauna occurs well represented in Antarctica is evidence that the connection between Antarctica and Africa must have been a broad as well as a close one. These two continents, now so completely separate, must in early Triassic time have been portions of a single land mass. If they had been connected by a long, narrow land bridge, similar, for instance, to the present Isthmus of Panama, there would almost surely have been a pronounced "filter effect," with a resultant exclusion of some faunal members from the one area or the other. Isthmian bridges, as we know from modern evidence, do not allow the unrestricted passage of land-living animals from one region to another; some animals are barred from crossing the bridge. But as between Africa and Antarctica there would seem to have been an almost complete flow of amphibians and reptiles back and forth—a freedom of migratory movement that could have taken place only if these two areas were, as already maintained, parts of a single continent.

The movements of amphibians and reptiles back and forth between Africa and Antarctica, thus bringing about an essential identity in the composition of the *Lystrosaurus* fauna as it occurs in the now two separate continents and thus indicating the close contiguity of these land masses during early Triassic times, were perhaps less extensive as they took place between Africa and what is now peninsular India. Or perhaps the fossil record is not so completely preserved in India. Whatever the explanation, the fact is that, although *Lystrosaurus* occurs plentifully in the Panchet beds of India, particularly in the area where those sediments are cut by the Damodar River, about 125 miles northwest of Calcutta, and although much of the *Lystrosaurus* from this region can be identified as *Lystrosaurus murrayi,* the species so typical of the African deposits, the remainder of the animal assemblage that is so characteristic in Africa is in India largely lacking. The thecodont *Proterosuchus,* or *Chasmatosaurus,* which we have seen in Africa, an almost invariable companion of *Lystrosaurus,* is present in India, as are some labyrinthodont amphibians (but of genera different from those found in Africa). And that is all, so far as the amphibians and reptiles

FIGURE 43. Early Triassic Gondwanaland. The scene is in South Africa, but it could as well be Antarctica or India. *Lystrosaurus,* the dicynodont reptile so ubiquitous in early Triassic Gondwanaland, is represented by a number of individuals, some feeding along the edges of the pond, others walking over the little hill toward the pond. The pond is also frequented by the armored thecodont reptile *Proterosuchus* or *Chasmatosaurus.* In the lower right corner are two advanced mammallike reptiles, *Thrinaxodon,* pictured as clothed in mammalian fur rather than reptilian scales. Such may have been the case. One is curled up, asleep; fossil skeletons frequently are found in this position. The other has its forefoot upon an ancestral lizard, *Prolacerta,* which it has killed. In the left foreground is *Procolophon,* a small, persistent cotylosaurian reptile.

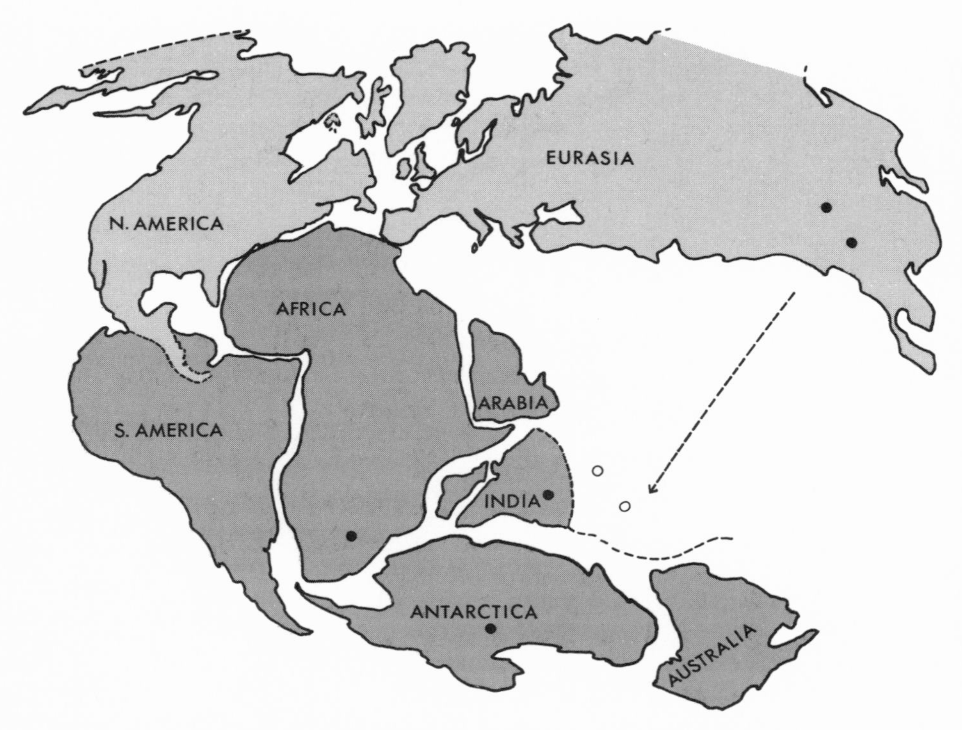

FIGURE 44. One view of early Triassic Pangaea. It has been suggested that much of eastern China may have been a part of Gondwanaland rather than of Eurasia. This would require the transfer of a part of Asia to a position contiguous to peninsula India. It is a very tentative hypothesis at the present time; for the paleontologist it is attractive because it brings all of the *Lystrosaurus* fauna localities into close association. Solid dots: known localities for the *Lystrosaurus* fauna. Circles: approximate positions of the Chinese *Lystrosaurus* fauna localities, if eastern China were a part of Gondwanaland.

are concerned. The deposits have produced the fossil remains of a primitive bony fish, of a crustacean, and of some unidentifiable plants. But there are no four-footed vertebrates other than *Lystrosaurus, Proterosuchus,* and the amphibians.

What does this mean? Does it mean that India was not so broadly connected with the African segment of Gondwanaland, or that there was some sort of a filter preventing various amphibians and reptiles from making the journey between an African center and the Indian region? If there was a filter effect, resulting in only a fraction of the *Lystrosaurus* fauna making its way from Africa into India, then this filter must have been something other than the usual narrow isthmian bridge. The geological evidence for a close relationship between India,

Africa, and Antarctica is too convincingly strong for us to imagine the present Indian peninsula as having been joined to the other sections of Gondwanaland by tenuous links. Perhaps the limited representation of the *Lystrosaurus* fauna in peninsular India is, as already suggested, simply the result of the accidents of preservation and collecting. At the present time it is hard to say.

The same questions arise with regard to the occurrence of *Lystrosaurus* in China, in Sinkiang and Shansi. According to one interpretation, we find here *Lystrosaurus* wandering far beyond the confines of the great southern continent, to invade the easternmost reaches of Laurasia. The fact that *Lystrosaurus* in China is accompanied by *Proterosuchus*, and by little else, makes plausible the idea of a long early Triassic migration of these animals from one great continent to another, with other members of the *Lystrosaurus* fauna "filtered out" along the way. Perhaps such was the case.

And perhaps not. For it is quite possible that China and even Indochina were integral parts of Gondwanaland, as we have seen. If such were the case, *Lystrosaurus* and its companions were members of a widely distributed fauna, confined to one large continental region.

Which brings us to the problem of the absence of the *Lystrosaurus* fauna in South America. All other evidence, including the evidence of fossils of a slightly later date, indicates that South America and Africa were intimately connected at the beginning of Mesozoic history, yet no traces of a *Lystrosaurus* fauna have been found in South America. Is this absence real? Perhaps it is; perhaps the various animals in the *Lystrosaurus* fauna were barred from South America by some factor unknown to us, even though the South American and African portions of Gondwanaland most probably were connected along a 2,000-mile front. It is quite reasonable to suppose, on the other hand, that the absence of the *Lystrosaurus* fauna in South America is one of the lacunae in the evidence resulting from our insufficient knowledge of a large and geologically incompletely explored continent. Let us be bold and predict that some day, perhaps in the not too distant future, a *Lystrosaurus* fauna *will* be found in the land to the south of the Amazon basin.

THE *CYNOGNATHUS* FAUNA AND ITS DISTRIBUTION

Certainly there is no doubting the fact that at a stage in the early Triassic slightly later than that represented by the *Lystrosaurus* fauna there was a very close faunal relationship between Africa and South

America. The *Cynognathus* zone of South Africa, coming directly above the *Lystrosaurus* zone, contains an abundant vertebrate fauna, dominated by therapsid reptiles, but containing as well some procolophonids —remnants of the ancestral reptilian cotylosaurs that had been prominent in Permian times, a few rhynchocephalians, a long-lasting group still represented in our modern world by the tuatara of New Zealand, two thecodonts, reptiles so typical of Triassic deposits, and a considerable array of labyrinthodont amphibians. In its general composition this fauna continues the labyrinthodont-therapsid complex that had so dominated the Gondwanaland tetrapod faunas of late Permian times; in its details it has its own distinctive features. It is rather more of an "upland" fauna than the preceding *Lystrosaurus* fauna, with an impressive array of carnivorous theriodont reptiles, large and small. The assemblage takes its name from the genus *Cynognathus,* briefly introduced earlier, a "wolflike" mammallike reptilian predator, some examples of which are actually as large as a large wolf. Accompanying this specialized carnivore, and various related meat eaters, is an interesting group of related mammallike theriodonts characterized by their broad cheek teeth. As mentioned in Chapter II, these so-called gomphodonts may have fulfilled the roles in early Mesozoic Gondwanaland that bears or raccoons do in our modern world; they may have been omnivorous carnivores, predators that fed upon a varied diet, animal and vegetable. Also quite characteristic of the *Cynognathus* fauna is the large two-tusked, or dicynodont, reptile *Kannemeyeria,* related in a general way to *Lystrosaurus,* but evidently adapted for life on high ground rather than in marshes and along watercourses.

As pointed out in Chapter II, portions of this fauna have been

FIGURE 45. *Cynognathus,* an advanced mammallike reptile characteristic of the lower Triassic sediments of South Africa and Argentina.

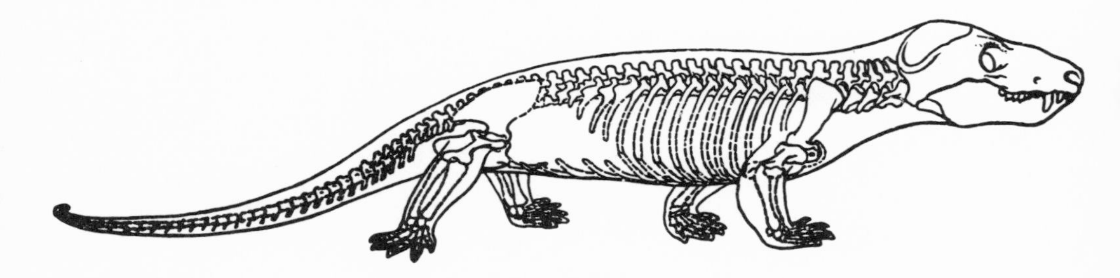

FIGURE 46. James Kitching searching for fossils in the Lower Triassic *Cynognathus* beds of South Africa.

found within the Lower Triassic Puesto Viejo Formation, in western Argentina. The correspondence is as yet not so complete as that between the earlier *Lystrosaurus* faunas in Africa and Antarctica, but it is sufficiently full to indicate the essential faunal, and therefore the physical, unity of Africa and South America. (It is this resemblance that has prompted the suggestion that some day the *Lystrosaurus* fauna may come to light in South America.)

Just as elements of the earlier *Lystrosaurus* fauna are found in peninsular India and China, probably as inhabitants of a Triassic Gondwanaland (or in China perhaps as wide-wandering intruders into Laurasia), so do elements of the *Cynognathus* fauna show similar patterns of distribution. The reptiles of the Yerrapalli Formation in India and of the Triassic beds of China are not so closely comparable to African *Cynognathus*-zone fossils as are the therapsids found in the Puesto Viejo beds of Argentina. Nevertheless relationships are obvious.

AUSTRALIA AND ANTARCTICA

The various lines of geological evidence showing a Paleozoic–early Mesozoic linkage of Australia with Antarctica, evidence such as the fit of continental outlines and the alignment of geosynclines, great structurally down-folded troughs of sediments, in recent years have been supported in a modest way by the discoveries of fossil amphibians and reptiles in Australia. In the Lower Triassic Blina Formation of northwestern Australia are labyrinthodont amphibians that reveal certain resemblances to amphibians from Antarctica, while similar amphibians and a few scattered reptiles, especially thecodonts and eosuchians, are found in Lower Triassic beds of Tasmania and Queensland. There is reason to think that the evidence will grow and become increasingly convincing in the years ahead.

A ROSTER OF LOWER TRIASSIC TETRAPODS IN GONDWANALAND

This review of Lower Triassic amphibians and reptiles demonstrates that these animals, so widely distributed in the present-day southern-hemisphere continents, as well as in India, were in a sense all of a kind. They were labyrinthodont amphibians of certain types, therapsid reptiles, both the highly modified dicynodonts, obviously adapted for living on plants, and the very mammallike theriodonts, obviously predators, as well as persisting cotylosaurs in the form of small, lizardlike procolophonids, other reptiles that were in effect the basic ancestors of the true lizards, and certain thecodont reptiles, from some of which in later ages were to arise the dinosaurs, crocodilians, flying reptiles, and birds. These active land-living tetrapods ranged widely across what are now southern continents and afford abundant and often detailed evidence, even down to the extensive spread of single genera, that the modern southern continents and peninsular India (and perhaps even China) probably were joined to form a supercontinent of Gondwanaland in early Mesozoic times, just as they had been during late Paleozoic history.

LOWER TRIASSIC AMPHIBIANS AND REPTILES OF LAURASIA

Suppose we now cross to the north of the equator, to view briefly the early Triassic amphibians and reptiles of Laurasia. The name "*Triassic*," based upon the Greek and Latin roots *trias*, meaning

FIGURE 47. The Lower Triassic Moenkopi beds of northern Arizona, in which, as in the Lower Triassic Buntsandstein of Germany, are found fairly numerous *Chirotherium* footprints, evidently made by thecodont reptiles, and occasional amphibian skulls. (Ancient Indians in the Arizona desert used the slablike Moenkopi rocks for construction of their pueblos.)

"three" or "threefold," was applied by early geologists who studied the rocks of this system exposed in central Europe. Here, in the type region of the Triassic, the threefold succession of the sediments is made up of the Lower Triassic, or Buntsandstein, at the bottom, consisting of brownish-red sandstones and shales containing amphibian and reptilian bones and footprints; the intermediate Middle Triassic, or Muschelkalk, consisting of thick limestones containing the fossils of marine organisms and representing a broad incursion of a shallow sea; and the Upper Triassic, or Keuper, consisting of red and variegated sandstones and clays in which are found the bones of land-living amphibians and reptiles. By all rights, Triassic rocks and fossils throughout the world should be compared with the threefold sequence in the type region. This is not always done.

The Buntsandstein, or Bunter, of central Europe is not abundantly fossiliferous. It contains occasional remains of labyrinthodont amphibians, especially those known as "capitosaurs," some scattered bones of reptiles, and in various localities the abundant footprints (showing a superficial but uncanny resemblance to a print that might be made by a human hand) of a reptile that has been called "*Chirotherium*." *Chirotherium* tracks are truly a feature of Lower Triassic geology in central Europe; at some places they are found by the hundreds and by the thousands, the fascinating records of the wanderings of ancient reptiles across ancient mud flats. Recent careful studies would seem to indicate that the footprints that have been called *Chirotherium* were made by thecodont reptiles.

This same association of amphibian bones with *Chirotherium* footprints, preserved in red sandstones and shales, is typical of the Lower Triassic Moenkopi Formation of Arizona and Utah. *Chirotherium* footprints are widely distributed in the American Southwest, and so similar are they to those of the European Buntsandstein that one might almost imagine hordes of thecodont reptiles, producers of the trackways, marching in a long procession between North America and central Europe. Of course, the significance of these trackways, as well as of the associated fossil bones, is that they betoken a broad, low-lying continental region in which the land-living tetrapods were widely distributed without the interposition of natural barriers. Again, as in the case of the Permo-Carboniferous amphibians and reptiles, the evidence of *Chirotherium* trackways and amphibian bones points to the continuing connection between northern Europe and North America as constituent elements of the ancient Laurasian continent.

The *Chirotherium*-labyrinthodont association or complex, if so it may be called, affords us a glimpse of an assemblage of amphibians and reptiles in the northern hemisphere, spreading latitudinally, and seemingly rather different in character from the Gondwanaland faunas, which also to a large extent spread latitudinally, but which were dominated by varied arrays of mammallike therapsid reptiles, instead of the dinosaur ancestors, the thecodonts.

Thus the early Triassic continents would seem to have shown something of a dichotomy in their land-living vertebrates, to match the geographical dichotomy of Gondwanaland and Laurasia. In the southern supercontinent the mammallike therapsid reptiles still reigned supreme, although there were the beginnings of a challenge from newcomers to the Mesozoic, the thecodonts and their relatives. In the northern supercontinent it would appear that the thecodonts were already the dominant reptiles and that they abundantly inhabited this part of the globe, perhaps almost to the exclusion of therapsids. In both regions there were varied labyrinthodont amphibians, carrying on the legacy of their Paleozoic past.

We have already seen that the confinement of faunas either to Gondwanaland or to Laurasia was no hard-and-fast matter. It would appear that the two supercontinents were connected into a Triassic Pangaea, and various active land-living vertebrates took advantage of this connection to move in north and south directions, as well as along east and west lines. We find that there were characteristically southern-hemisphere wanderers in northern Russia. Here there are dicynodonts, related to the South African *Kannemeyeria* of the *Cynognathus* zone, associated with thecodont reptiles, and the ubiquitous Lower Triassic labyrinthodont amphibians. And so, as is usually the case, our view of the distribution of Lower Triassic reptiles does not provide us with a clear-cut or simple picture.

Nonetheless, it is probably valid to say that in early Triassic time,

FIGURE 48. The association of Lower Triassic thecodont reptile footprints (*Chirotherium*) and amphibian skulls (capitosaurs) in Laurasia. These same fossils are found in the Moenkopi Formation of Arizona and the Buntsandstein of Germany—evidences of land-living vertebrates that wandered across the continuous lands of a Laurasian continent. In this figure continental-shelf areas are indicated by oblique lines, overlaps by black areas, and gaps between the fitted continents by wavy lines.

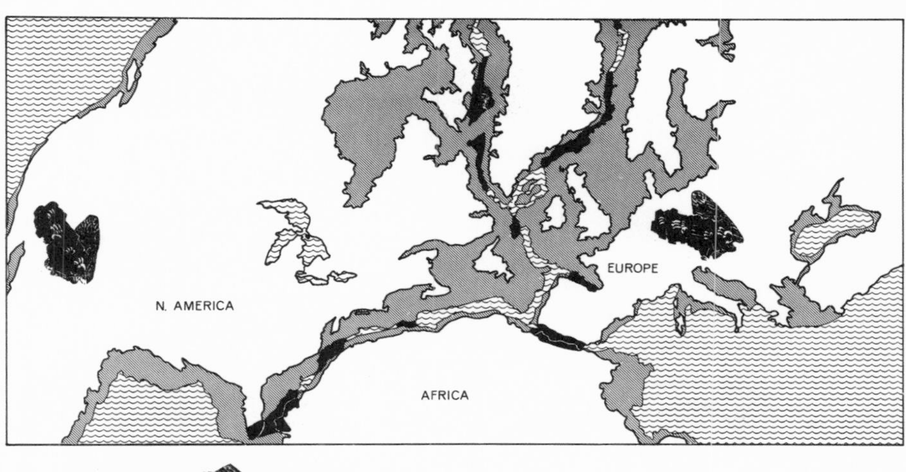

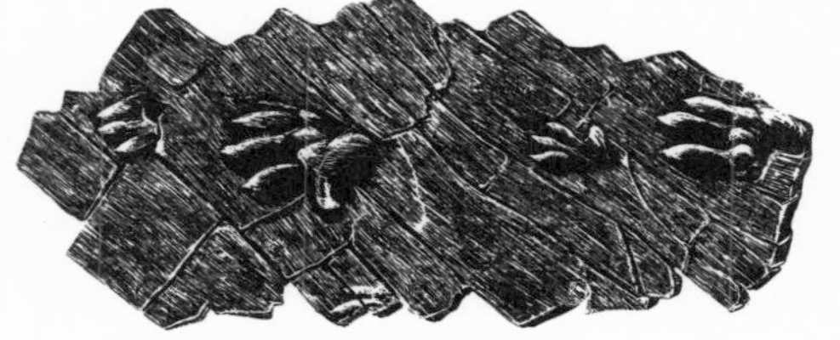

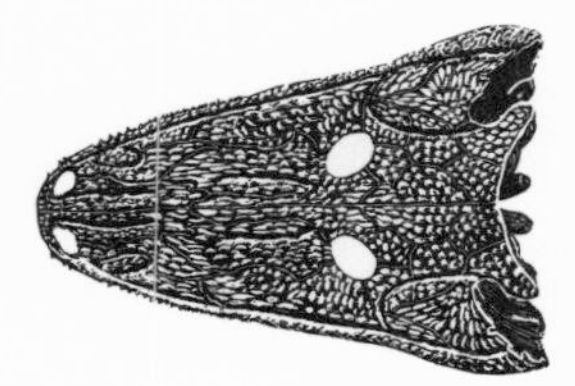

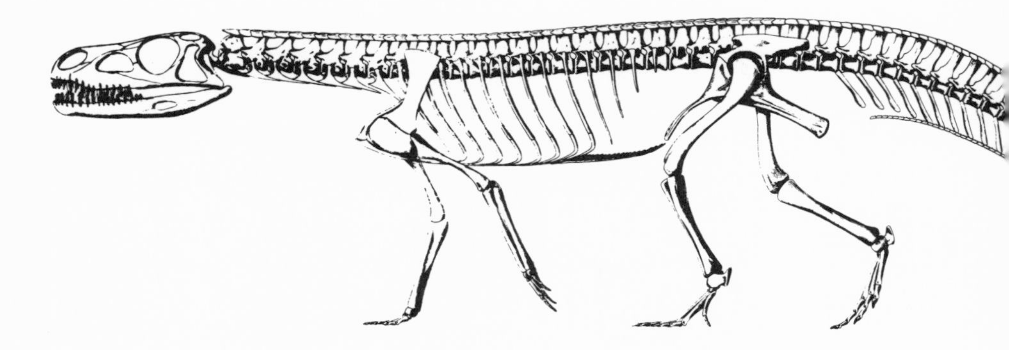

as in the days of late Paleozoic history, there were on the earth two vast regions for the development of land life—Gondwanaland and Laurasia. These regions certainly were not mutually exclusive; animals passed from the one to the other as opportunities allowed them to do so. As we have seen, Pangaea was for long ages a connected world, yet generally speaking Gondwanaland was a land in which mammallike therapsid reptiles were still the dominant tetrapods, as they had been in the preceding Paleozoic days, while generally speaking Laurasia was a land in which the protodinosaurian thecodont reptiles had already become established in great numbers.

THECODONTS—KEY REPTILES OF THE TRIASSIC

These thecodonts, among which had become established the anatomical characters that by and large were to determine the success of the dinosaurs—namely, a lightly constructed but well-articulated skull, powerful jaw muscles, a supple skeleton, and long, strong limbs—were harbingers of the future. Their descendants would inherit the earth during the next two geologic periods, and even during the last of the Triassic period.

Represented in the Lower Triassic sediments of Gondwanaland by a few scattered genera, such as *Proterosuchus*, the constant companion of *Lystrosaurus*, and in Laurasia mainly by the prolific trackways of *Chirotherium*, the thecodonts become much more evident in deposits of middle Triassic age. This record of thecodonts is, however, to be found largely in Gondwanaland, because in Laurasia, and especially in the type Triassic region of Europe, the Middle Triassic was an interlude so far as land-living vertebrates are concerned. This was the time of the Muschelkalk incursion, when a shallow sea covered much of central

FIGURE 49. The skeleton of *Ticinosuchus,* a Triassic thecodont reptile of the type that might have made *Chirotherium* tracks.

Europe, when this region was a marine basin in which were deposited thick limestones and shales containing an abundant record of seashells, as well as of various marine reptiles, which in those days abounded on the ocean surface. In Britain the Muschelkalk is absent, and in North America there seems to be no record of continental Middle Triassic. Laurasia obviously is not the place for the student of Middle Triassic land-living tetrapods.

Within recent years, however, there has come to light a very nice thecodont skeleton in the Middle Triassic deposits of Switzerland. This skeleton, named *Ticinosuchus,* provides us with a detailed glimpse of an undoubted Middle Triassic thecodont, a sizable reptile, perhaps eight feet in length, with rather long, slender limbs carried well beneath the body; a large, open skull, the jaws being set with sharp, pointed teeth; a long tail; and armor scutes down the back. Incidentally the feet of *Ticinosuchus* could very well have made the tracks that have been christened *Chirotherium.*

Reptiles very similar to *Ticinosuchus* have been found in the Santa Maria Formation of southern Brazil, and the Ischigualasto Formation of western Argentina, in both instances associated with early dinosaurs, which latter are often considered as of late Triassic age, at the earliest. Whatever may be the exact age of these beds (and this is a rather academic question in the present context), the fossils in them do have a definite Gondwanaland cast, since with the thecodonts and the primitive dinosaurs are associated dicynodonts and very mammallike theriodonts.

It would appear that during the years of the middle Triassic and on into the beginning of late Triassic time there was still a flow of tetrapods, especially the mammallike reptiles, back and forth between Africa and South America. But, as indicated by the thecodont reptiles and the dinosaurs, there evidently was some interchange between South America and Europe, probably through northern Africa.

So it would seem that the land-living tetrapods of the middle Triassic perhaps were still somewhat distinctive in the two supercontinents, as they may have been during early Triassic time. Gondwanaland remained a land of mammallike reptiles, and Laurasia a land of thecodonts, but there was an ever-increasing blurring of the faunal lines of demarcation between the two regions. This was perhaps the result in part of a southern-hemisphere increase of thecodonts—because of incursions from the north. Such a development boded no good for the varied mammallike reptiles, since the thecodonts and their descendants increased at the expense of the therapsids. Late Triassic history belonged to the thecodonts and dinosaurs while the therapsids or mammallike reptiles dwindled to mere remnants of their former numbers and variety.

THE COSMOPOLITAN UPPER TRIASSIC DINOSAURS AND THEIR CONTEMPORARIES

This is strikingly shown in the Upper Triassic deposits of southern Africa, in the long slopes of the Red Beds and in the stunning white and red cliffs of the Cave Sandstones fringing the high, volcanic Drakensberg, which forms the core of Lesotho, the inland kingdom set within the confines of the Republic of South Africa. In these sediments there have been found the bones of various fossil reptiles, and the reptiles are predominantly thecodonts, ancestral crocodilians and many primitive dinosaurs, with the mammallike reptiles represented by a few small, very specialized stragglers.

Such a pattern of thecodont-dinosaur dominance is not peculiar to southern Africa; it is typical of Upper Triassic rocks throughout the world. For example, in South America, where the Lower and Middle Triassic tetrapod faunas (so far as they are known) are strikingly parallel to those of Africa, the Upper Triassic tetrapods, as represented especially in the Colorados Formation of western Argentina, show an abundance of armored thecodonts and of primitive dinosaurs, associated with the strangely adapted rhynchosaurs, so characteristic of many Middle and Upper Triassic deposits, and with a few persistent mammallike dicynodonts.

A similar association of reptiles is seen in the Upper Triassic

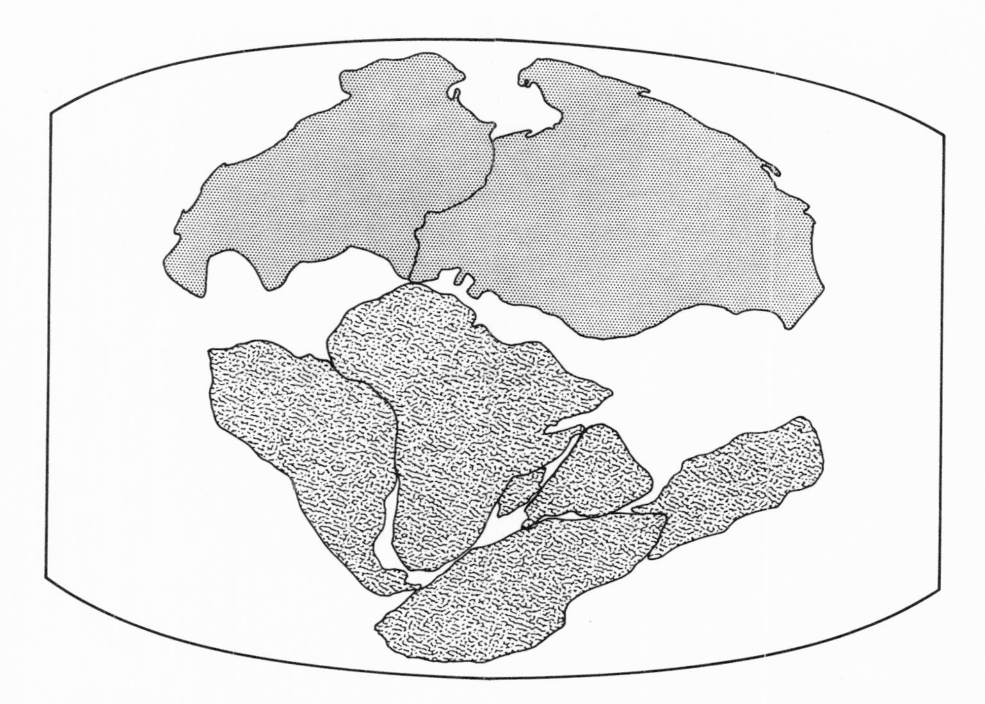

FIGURE 50. A reconstruction of Pangaea during Triassic times. India and Antarctica-Australia are shown as connected with Africa, as seems necessary to account for Triassic land-living tetrapods of African relationships in these regions.

Maleri Formation of central India. The thecodonts and the primitive dinosaurs of the Maleri beds reflect faunal conditions that were worldwide in the final days of Triassic history; thecodonts and dinosaurs were ubiquitous. The rhynchosaurs of India are especially interesting in that they would seem to show close relationships to the rhynchosaurs of South America, thus indicating connections across a broad latitudinal range, from the western border of the South American block across the width of Africa and through much of the length of the Indian peninsula. Of particular interest, however, is the presence in the Maleri beds of a large labyrinthodont amphibian, *Metoposaurus*, identical with similar amphibians found in Germany, and across almost the full width of North America, from New Jersey and Pennsylvania to New Mexico and Arizona. And associated with this big amphibian in the Maleri beds are phytosaurs—long-snouted, crocodilianlike thecodonts, very close to if not identical with phytosaurs found in Germany and in the same parts of the United States where the metoposaurian amphibians occur. Certainly there was an open avenue along extraordinarily long lines for

FIGURE 51. Upper Triassic tetrapods. M—metoposaur, a labyrinthodont amphibian. P—phytosaur, a thecodont reptile. T—tritylodont, an advanced mammallike reptile. S—saurischian dinosaur. D—dicynodont reptile. R—rhynchosaur.

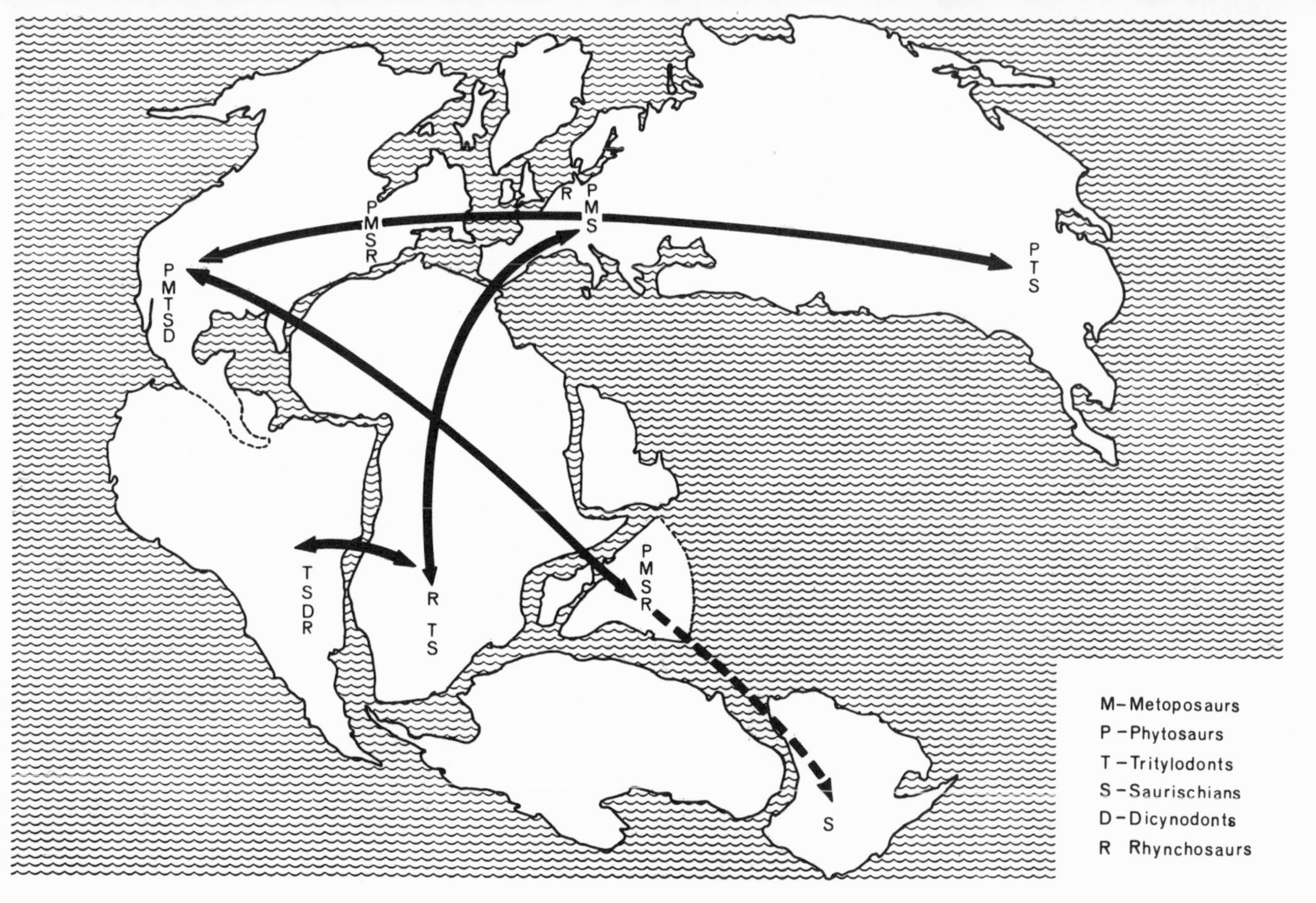

FIGURE 52. Pangaea in late Triassic time, showing the inferred migrations of the tetrapods, as indicated by letters, pictured in figure 51.

FIGURE 53. At Ghost Ranch, northwest of Santa Fe, New Mexico, is a magnificent sequence of Mesozoic sediments. The Upper Triassic Chinle Formation forms the rounded hills at the base of the cliffs. From these brilliant red sandy clays have come fine specimens of phytosaurs, armored thecodont reptiles, and primitive saurischian dinosaurs. (The vertical cliffs are formed of Jurassic rocks, as are the long slopes above them—these being the Morrison beds, containing large dinosaurs. On the skyline are Cretaceous rocks.)

the diffusion of those amphibians and phytosaurs, characteristic of the late Triassic of Laurasia, into the Indian section of Gondwanaland. India would thus seem to have occupied, in late Triassic time, the position of a faunal crossroad, with connections to other parts of Gondwanaland, some of which were perhaps 5,000 miles distant, and with equally accessible connections to areas in Laurasia that were at least 8,000 miles away. Here we see a nice example in the development of cosmopolitan faunas of land-living tetrapods, which were to so characterize this and later spans of earth history.

The Upper Triassic sediments of the Lufeng Formation, in Yunnan, China, have yielded the inevitable thecodonts and dinosaurs that one may expect in deposits of this age. One of the dinosaurs, at least, is of particular significance—namely, the genus *Lufengosaurus*. This interesting dinosaur—a rather large prosauropod (the prosauropods being more or less ancestral to the huge sauropod dinosaurs of Jurassic and Cretaceous age), 20 feet or more in length, with a massive body borne upon heavy hind limbs, the comparatively short forelimbs being capable in life of some support, but also obviously useful for grasping and food gathering, with a long, sinuous neck surmounted by a small skull in which the numerous, small spatulate teeth were evidently adapted for the eating of plants—is very close indeed to the dinosaur *Plateosaurus,* from the Upper Triassic Keuper beds of southern Germany. It is likewise very closely related to the genus *Plateosauravus,* from the Red Beds of South Africa. So it is that these three dinosaurs, so very similar to one another, yet so widely separated on the earth today, give us striking evidence as to the worldwide distributions of many tetrapods during the final phases of Triassic history.

The cosmopolitan nature of late Triassic faunas is emphasized by still another reptile in the Lufeng deposits, the very highly adapted mammallike therapsid *Bienotherium.* This reptile is a tritylodont, belonging to a group the name of which is derived from the genus *Tritylodon,* found in the South African Red Beds. *Bienotherium* is very closely related to *Tritylodon;* tritylodonts of this same aspect have been found in the Kayenta Formation of Monument Valley, where the states of Arizona and Utah meet in a desert region of breathtaking magnificence. Also, tritylodonts have recently been discovered in the Upper Triassic of Argentina. The tritylodonts, among the last of the mammallike or therapsid reptiles, were animals of moderate size, perhaps comparable in this respect to a small spaniel. The body in *Tritylodon* and *Bienotherium* is robust, supported by strong, rather short limbs. The skull shows specializations that remind one of a large rodent; it is

FIGURE 54. Skull and jaw of a tritylodont—a very advanced mammallike reptile of late Triassic age. Fossils of this type are known from Upper Triassic beds in South Africa, Argentina, China, and Arizona—an indication that these vertebrates were distributed widely throughout Pangaea.

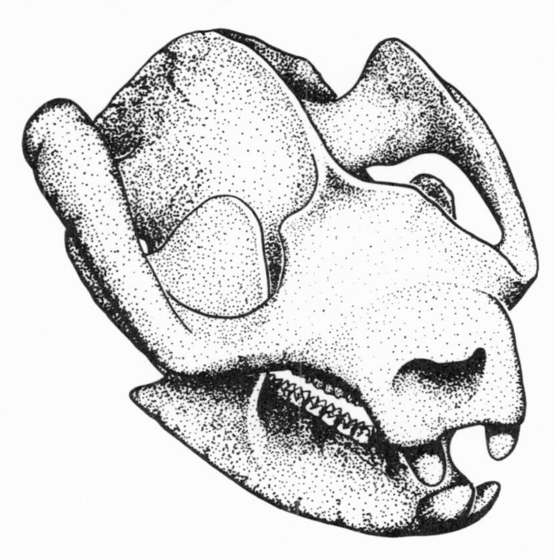

FIGURE 55. *Coelophysis*, a small saurischian dinosaur from the Upper Triassic Chinle Formation at Ghost Ranch, New Mexico. A dinosaur remarkably close to *Coelophysis* has been found in Upper Triassic sediments in southern Rhodesia.

elongated and rather low, there are large peglike teeth in the very front of the skull and lower jaws, separated by a considerable gap, or diastema, from the complex cheek teeth, in which latter there are longitudinal rows of sharp cusps. It would appear that the tritylodonts probably fed upon plants, plucking their food with the large anterior teeth and grinding it in the rasplike mill of the cheek teeth by fore and aft movements of the lower jaws. These animals reinforce the evidence of the dinosaurs, to show that many of the late Triassic reptiles roamed far and wide across Laurasia and Gondwanaland.

The classic Upper Triassic tetrapod fauna is the assemblage of amphibians and reptiles found in the Keuper deposits of the type Triassic of central Europe, notably in southern Germany, where the lovely wooded hills and the fertile valleys of Württemberg have been the locale for many a pleasant and fruitful paleontological dig during the past two centuries. Here are found large labyrinthodont amphibians, including the big metoposaurs that already have been mentioned as typical of the Upper Triassic of North America and India. Here also are found primitive dinosaurs, among which may be mentioned *Plateosaurus,* already referred to, armored thecodonts and crocodilianlike phytosaurs (also thecodonts), and ancestral turtles and lizards.

This same general association of amphibians and reptiles is found

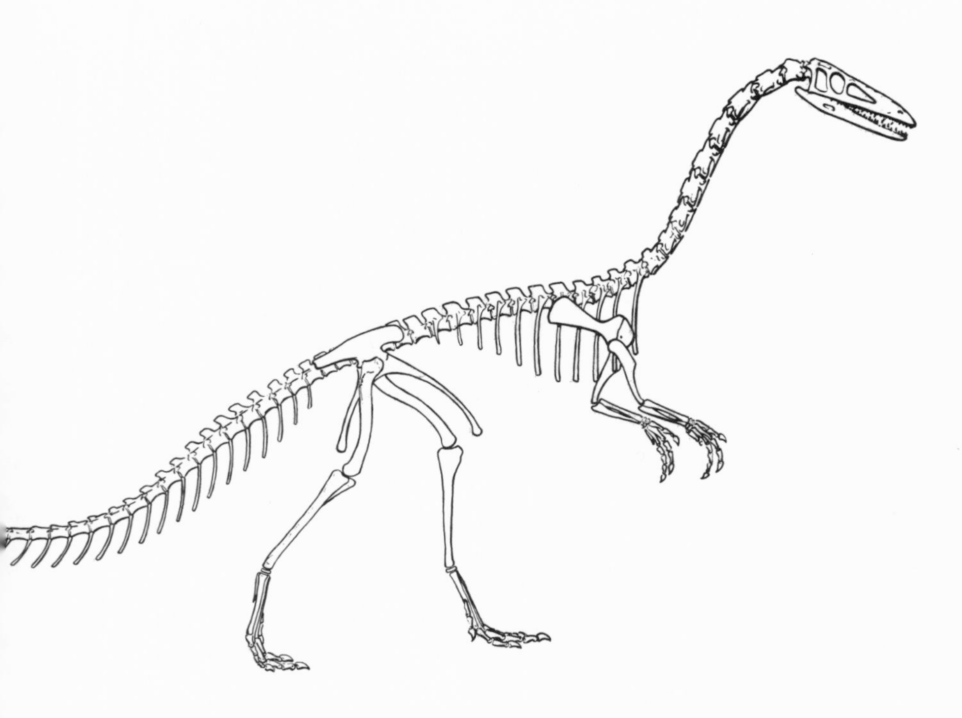

in North America, in the Newark sediments of the eastern seaboard from Nova Scotia to the Carolinas, in the Dockum beds of Texas and eastern New Mexico, and in the Chinle, Wingate, Moenave, and Kayenta formations (from bottom to top) in the southwestern deserts of New Mexico, Arizona, and Utah. The resemblances of the amphibians and reptiles of the North American continent to those of central Europe are so close that we are justified in assuming the two regions to have been part of one great land mass, as had been the case in earlier Triassic and in Permo-Carboniferous times. In North America, however, there are several reptiles that extend the relationships of late Triassic tetrapods in this region beyond their European ties. There are, of course, the tritylodonts, which, as we have seen, widen the horizons of reptilian distributions from North America to western China and to South Africa. And quite recently there has been discovered in Rhodesia a small, primitive dinosaur, given the name of *Syntarsus,* which shows remarkably close resemblances to the North American dinosaur *Coelophysis,* from the Chinle and Newark sediments. Both of these dinosaurs are small, lightly built reptiles with hollow bones. They are about eight or ten feet in length, which may seem large, but owing to the slender build of the skeletons it is obvious that in life these dinosaurs probably weighed less than 100 pounds. The neck and tail are very long, the neck carrying a small, elongated skull, the jaws of which are set with sharp teeth. The hind limbs are long and birdlike; the forelimbs are short and obviously were not used for locomotion, but rather for grasping. Evidently these dinosaurs were fleet, active reptiles, preying upon other small reptiles and amphibians. Continuing, there are true ancestral crocodiles in the Triassic beds of western North America, as well as in the Red Beds and the Cave Sandstones of South Africa. Finally, in the Chinle Formation of Arizona there is a gigantic mammallike dicynodont, *Placerias,* perhaps the last of these strange tusked therapsid reptiles. Giant dicynodonts occur in the Ischigualasto and Los Colorados formations of Argentina and in the Santa Maria beds of Brazil. The large dicynodonts in South America evidently are the last survivors on that continent of a group that enjoyed its evolutionary development in Gondwanaland; *Placerias* would seem to represent an invader that pushed into Laurasia during late Triassic time.

Such is the view, as derived from the fossil evidence, of tetrapod distributions on the ancient continents during the progression of Triassic time. The crosscurrents of tetrapod distribution that took place during late Triassic history would seem to have brought a new degree of unity to the land-living vertebrates of the world. As we have seen, there probably never was anything like a complete faunal distinction between Gondwanaland and Laurasia in early Triassic time, even as there had not been during late Permian history. Nevertheless, in spite of crossings back and forth, from north to south and from south to north, Gondwanaland and Laurasia would appear to have been in late Paleozoic and early Triassic times two great zoogeographical realms in terms of land-living vertebrates—the one dominated by mammallike therapsid reptiles, the other by, in turn, pelycosaurs (the ancestors of the therapsids) and thecodonts (the ancestors of the dinosaurs). Then with the advent of middle Triassic events, the lines between the two regions began to be progressively blurred—a process that continued into the late Triassic, when land-living vertebrate faunas became very much alike throughout the world.

This change in the nature of the faunas was in essence a change in tetrapod dominance developing during the transition from early to late Triassic time. The Triassic period of transition, as it has here been called, saw the displacement of the old and long-established mammallike therapsids, which had enjoyed such a long history of radiation in Gondwanaland, by the new thecodonts and, after them, the dinosaurs and crocodilians that from late Triassic time until the end of the

Cretaceous were to be supreme on the surfaces of the changing continents.

The displacements of old groups of vertebrates by vigorous newcomers were events taking place at the end of Triassic time. They marked the beginning of a new age of life, which was to continue through the long years of Jurassic and Cretaceous history. At the same time there began a long sequence of continental movements that were to affect profoundly the development of land-living vertebrates in the years to come.

V. THE BREAKUP OF THE ANCIENT WORLD

EXTINCTIONS AND REPLACEMENTS DURING THE TRIASSIC-JURASSIC TRANSITION

The changing of the guard among the backboned rulers of the continents that took place between the early and the final days of Triassic time was followed, at the close of this period of earth history, by an almost complete disappearance of those tetrapods which formerly had been supreme across much of the earth's surface, particularly across ancient Gondwanaland. The crocodilians and particularly the dinosaurs, abetted by early turtles, ancestral lizards, and ancient frogs, had by the end of the Triassic displaced the once-ubiquitous labyrinthodont amphibians and the widely distributed mammallike reptiles, as well as other Paleozoic reptilian holdovers, including the persisting cotylosaurs, which until the appearance of the ancestral lizards had played the ecological role of lizards in the Triassic scene. There was no room on the earth for these descendants of ancient evolutionary lines—and they disappeared. Almost so, but not quite; one small group of mammallike reptiles, the highly adapted tritylodonts, with their strange rodentlike skulls and teeth, carried over into the opening phases of Jurassic history before becoming extinct. Then they, too, succumbed.

It should be added that among the tetrapod casualties marking the transition from Triassic to Jurassic time there must be included one order among the newly evolved reptiles—the thecodonts. It will be recalled that the active and varied thecodonts were in many respects the forerunners of the transition among tetrapod faunas, which took place between the beginning and the end of Triassic history. The thecodonts appeared essentially with the advent of Triassic time, to herald the evolutionary growth of the ruling reptiles of the Mesozoic era. They were highly successful during the course of Triassic history, not only in their own right, but also as progenitors of those tetrapods destined to rule the middle and late Mesozoic world—the crocodilians, dinosaurs, flying reptiles, and birds. By late Triassic time the crocodilians and the dinosaurs had become firmly established and widely distributed; they carried on while the last of the thecodonts, descendants of ancestors which had given rise to the Mesozoic rulers, declined and finally became extinct. But not before some of them had evolved in such ways that would lead in the direction of flight. It is obvious that the flying reptiles, or pterosaurs, and the birds are of thecodont ancestry, although there are no records in the rocks of pterosaurs and birds as such until we reach sediments of Jurassic age.

Another group of tetrapods involved in this change from the old to the new, a group that has not previously been mentioned because it was not previously present in the Triassic scene, was that of the first mammals. The first mammals, small and insignificant in appearance, had made their appearance during the later phases of Triassic history as undoubted descendants from certain mammallike reptiles. The transition across the threshold from reptile to mammal, as preserved in the fossil record, is so gradual—so smoothly accomplished in the evolutionary sense—that it is a moot point as to where the reptiles leave off and the mammals begin. The evolutionary change from the one great class of vertebrates to the other involves so small a step that it is necessary to rely upon a very esoteric detail of anatomy—the nature of the jaw articulation—to determine the separation between reptile and mammal.

Of course, we are forced to deal with the fossilized bones in this view of the crucial reptilian-mammalian transition. Perhaps if we could see the whole animals the distinction between the two classes might be easier. Perhaps not. The mammals introduced revolutionary morphological and physiological innovations into the vertebrate world, such as a constant high body temperature, insulating hair, live birth, the suckling of the young on milk, an enlarged brain, and a diaphragm for more efficient breathing, to mention a few. But these changes came gradually, and there is good reason to think that many of the changes had already taken place among the highly advanced mammallike reptiles. Moreover, we know that among the most primitive living mammals, the monotremes—represented by the platypus and the echidna of Australia—some of the primitive reptilian traits are still retained, such as the birth of the young from eggs. So it is evident that the origin of mammals from reptilian ancestors was a gradual affair.

The fossil evidence does indicate, however, that the final step in this long-continuing process had occurred by late Triassic time. And from that time until the close of the Cretaceous period the mammals played increasingly larger roles, although seemingly not particularly important ones, in the composition of tetrapod faunas. Their time of evolutionary dominance was to come later. Nonetheless, they were present, and perhaps they contributed in some degree to the disappearance of the old-line tetrapods at the end of Triassic time. Thus, in a way, some of the mammallike reptiles, those that were the direct ancestors of the first mammals, evolved themselves out of existence.

Extinction is an old and often repeated phenomenon in the story of life through the ages and one that, despite all efforts at interpretation, is ill-understood. Extinction is a necessary and, in one sense, a creative process in the course of evolution, for without the extinction of species as a corollary to the origin of species, there would be no advancement in the development of life. The organic world would be stagnant. It is essential that the old make way for the new, in order that the new may have living space, and with such space the opportunity to evolve—to become adapted to a changing world. Of course, the old does not passively make way for the new; rather the new crowds out the old and the less adaptable forms of life. So it was at the close of Triassic time; the progressive reptiles crowded out many land-living vertebrates, which for so long had enjoyed dominance on the lands of the earth.

The displacement of those reptiles and amphibians that became extinct with the close of the Triassic might more properly be thought of as a replacement by newer and more successful types of vertebrates. Just why these newer animals should have been "more successful" than their predecessors is a hard question; the fact is that they were, and they prevailed. Undoubtedly this phenomenon of extinction and replacement during the transition from Triassic to Jurassic time was but one among the many changes that took place as the ancient worlds of Gondwanaland and Laurasia began to break apart. It was a time of great changes affecting the lands and the life of the earth.

The disruption of Pangaea was a profound and long-enduring event in geological history and will be considered in the pages that follow. But before we come to this large and complex subject let us summarize the manner in which the land-living tetrapods that died out at the end of the Triassic period were replaced. This may be presented in tabular form:

EXTINCTION AND REPLACEMENT AMONG LAND-LIVING TETRAPODS AT THE END OF TRIASSIC TIME

Groups that became extinct at the end of the Triassic	Their environment	Their diet	Replaced by
Labyrinthodonts	Rivers, lakes	Fishes	Crocodilians
Thecodonts, phytosaurs	Rivers, lakes	Fishes and other tetrapods	Crocodilians
Protorosaurs	Edges of ponds, thickets, uplands	Insects, small tetrapods	Lizards and frogs
Cotylosaurs, procolophonids	Thickets, rocks	Plants	Lizards
Small thecodonts	Uplands	Insects, small tetrapods	Small dinosaurs
Armored thecodonts	Uplands	Plants	Plated dinosaurs
Dicynodonts	Uplands	Plants	Herbivorous dinosaurs
Rhynchosaurs*	Uplands	Plants	Herbivorous dinosaurs
Large theriodonts	Uplands	Various tetrapods	Carnivorous dinosaurs
Small theriodonts	Uplands	Insects, small tetrapods	Ancestral mammals

* Although the rhynchosaurs, typical of Triassic faunas, became extinct, the reptilian order to which they belonged, the **Rhynchocephalia**, continued.

Needless to say, a listing such as this contains some highly conjectural elements. We can make inferences as to the habitats in which lived the various tetrapods on this list, by virtue of the differing adaptations in their skulls and skeletons, as well as the nature of the sediments in which their fossilized remains are preserved. We can make inferences as to what they ate, by virtue of the structure of their jaws and teeth. But these are largely inferences, for we are looking at the petrified remnants of animals removed from us by almost 200 million years of earth history. Nevertheless, the inferences may be reasonable ones, and they may help us to understand the nature of life in a changing world.

THE CHANGING WORLD OF LATE TRIASSIC TIME

The world *was* changing—of that there can be little doubt. The end of the Triassic very probably marks the beginning of the end of Gondwanaland and Laurasia, with the inauguration of physical changes of overweening importance that were to determine the evolutionary directions that would take place among the earth's living inhabitants. Why is the end of the Triassic period chosen as the time when the revolution in the configuration of the land masses on the earth had its beginning? This is a pertinent question and one for which there is no precise answer. Many authorities have suggested other dates, most of them at later stages in earth history, for the initiation of the breakup of Pangaea. These authorities may be right, yet in spite of arguments in favor of later dates for this important event, there is good reason to think that perhaps the transition from the Triassic into the Jurassic period of earth history is the time when the rifting of Gondwanaland and Laurasia began.

THE FRAGMENTATION OF PANGAEA

Of course, there were the extinctions and replacements among the land-living vertebrates that have already been pointed out. These may very well have taken place on a static and stable earth, but there is something to be said for the supposition that the faunal changes whereby the old and long-established evolutionary lines of land-living tetrapods vanished in the face of pressures from newly developed groups of animals which had arisen during the course of Triassic history were among the many concomitant events marking physical changes that were to set the Jurassic and Cretaceous world apart from the world of earlier times. Moreover, there are other good reasons for thinking that perhaps the great breakup occurred at this time. Quite recently Drs. R. S. Dietz and J. C. Holden, who have given much attention to this problem, have suggested that the widespread eruptions of volcanic rocks in late Triassic times signaled the beginning of the end of Pangaea: "We take the immediate prelude to the breakup of Pangaea to be the first large outpourings of basaltic rock along the continental margins being established by rifting."

If one accepts Gondwanaland and Laurasia, connected to form a greater Pangaea, then one perforce must accept their dissolution by rifting into several separate masses, and the drifting of the various resulting continental blocks to their present positions. The theories of Gondwanaland, Laurasia, and Pangaea and of continental drift go together.

Although the two theories are complementary, the lines of evidence to support them are of different sorts. As we have seen, the remarkable "fits" between the several continental blocks, in conjunction with geological features that connect them, supported by the distributions of land-living vertebrates (among other fossil evidence) of late Paleozoic and early Mesozoic age, give substance of a very solid sort to the concept of Gondwanaland and Laurasia. The breakup of the ancient continents is a problem for which other evidence is needed.

Wegener and Du Toit could evoke no seemingly valid mechanism to explain the breaking up of the supercontinents and the subsequent drifting of the continental fragments away from each other. This lack of a good mechanism to account for continental drift was a serious stumbling block for the theory, especially in the minds of structural geologists and geophysicists, who are concerned with earth forces.

So the matter stood, until very recent years. Then, in the past ten years or so, there has been a flood of new evidence, of kinds never envisaged by Wegener or Du Toit or their immediate followers, that has drastically altered the situation. This evidence, such as paleomagnetism, magnetic-field reversals, sea-floor spreading, and plate tectonics, added to the previous correspondence of continental outlines, the distribution of fossils and of modern organisms, and the similarities of geological features in various continents, has shifted the theory of continental drift from the realm of a collection of possibilities to one of an array of probabilities. Continental drift is assuming the aspect of reality.

PALEOMAGNETISM AND CONTINENTAL DRIFT

Perhaps the reader is appalled or quite put off by the recital of seemingly esoteric geological terms in the previous paragraph. They are not so formidable as they appear to be, and a few words of explanation may be useful.

Paleomagnetism has already been briefly outlined in Chapter II. But to repeat, when hot volcanic rocks are cooled, the magnetic particles within the rocks become aligned in relation to the magnetic poles of the earth. These magnetic lines are in a sense "fossilized," or "frozen," and they remain in their frozen attitudes through all subsequent ages. Thus if the paleomagnetism of a volcanic rock is now in an east-to-west direction, it is to be supposed that the rock has been rotated from an original north-and-south direction—the direction in which it presumably was oriented when cooling took place. (This method may also be applied to certain sedimentary rocks.) Furthermore, if evidence shows that the rock has not been disturbed within the framework of the continental block in which it rests, then it is to be supposed that the continental block has been rotated through time. By studying paleomagnetic directions in the several modern continents in rocks of different geological ages, it is possible to reconstruct the attitudes of the continents to ancient paleomagnetic poles during past geologic periods. Then, assuming that the paleomagnetic poles were closely related to the poles of earth rotation, as are the modern magnetic poles, it is possible to reconstruct the positions of ancient continents with relation to the poles of rotation.

Recent studies have indicated that the earth's fields of magnetism have been suddenly and repeatedly reversed during geological time. This is the phenomenon of magnetic reversals, characterized by alternating periods of "normal" and "reversed" magnetism.

PLATE TECTONICS AND CONTINENTAL DRIFT

Plate tectonics is the modern concept that the surface of the earth is divided into a number of huge plates, each including not only a continental mass, but portions of oceanic basins as well. These plates are and have been constantly moving in relation to each other, such movement being expressed by sea-floor spreading, which is the passive movement of oceanic crust away from the midoceanic ridges—with which we will shortly become acquainted.

Such is the mechanism of continental drift. There is much very convincing evidence to support this very grand concept of the earth.

The explanation of the mechanism of continental drift, as based upon the concept of plate tectonics and sea-floor spreading, has been one of the very exciting developments in theoretical geology during the past few years. At last, a probable answer has been found to the question that for so long has perplexed geological theorists, the question for which Wegener and Du Toit had no answer.

As the result of extensive oceanographic surveys during the years since World War II, many of them conducted in vessels sent out by the Lamont Geological Observatory of Columbia University, it has become plainly evident that there are immensely long ridges running along the floors of the oceans between the continents. One of these is the mid-Atlantic ridge, extending north and south about equidistant between Europe and North America and between Africa and South America. Another ridge runs parallel to the west coast of Middle America, veers westward to occupy the ocean floor some thousands of miles to the west of South America, then curves through the South Pacific, to the south of Australia, and up through the Indian Ocean to the Persian Gulf. A branch extends around the tip of Africa, to join the South Atlantic ridge.

These ridges, though essentially continuous, are broken by numerous faults transverse to their axes, so that the midlines of the ridges are frequently offset, either to the right or to the left. Furthermore, earthquakes are quite obviously associated with the midoceanic ridges.

The pattern of the midoceanic ridges began to become apparent in the late 1950s, when the data of the various oceanographic surveys were collected and collated. At about this same time, as a result of the accumulation of many deep-sea cores made from the survey ships, it also became apparent that the oceanic basins are relatively young. Thus, in the several oceans no sediments older than Cretaceous or late Jurassic age were found as the result of coring, and much of the cored materials were of Cenozoic age.

About 1960 Professor Harry H. Hess of Princeton University suggested that there might be movement in the floor of the ocean. He envisioned molten magmas rising from the depths, along the axes of the midoceanic ridges, spreading laterally across the floors of the oceans, away from the ridges. To compensate for this addition of rock materials to the crust, the ocean floor presumably disappeared downward into trenches frequently located along the edges of the continents.

As we have seen, when molten rocks cool the iron particles in these rocks are aligned along lines of force directed toward the earth's magnetic poles, so that the magnetic directions, either normal or reversed, are "frozen" into the rocks and preserved, no matter how the rocks may be moved and shifted by earth forces during subsequent geologic ages. With this phenomenon as a basis, F. J. Vine and D. H. Matthews, of Cambridge University, proposed in 1963 that as molten rocks welled up along the axes of the midoceanic ridges, they would, as they cooled, preserve a record of the earth's magnetic field at the time of their

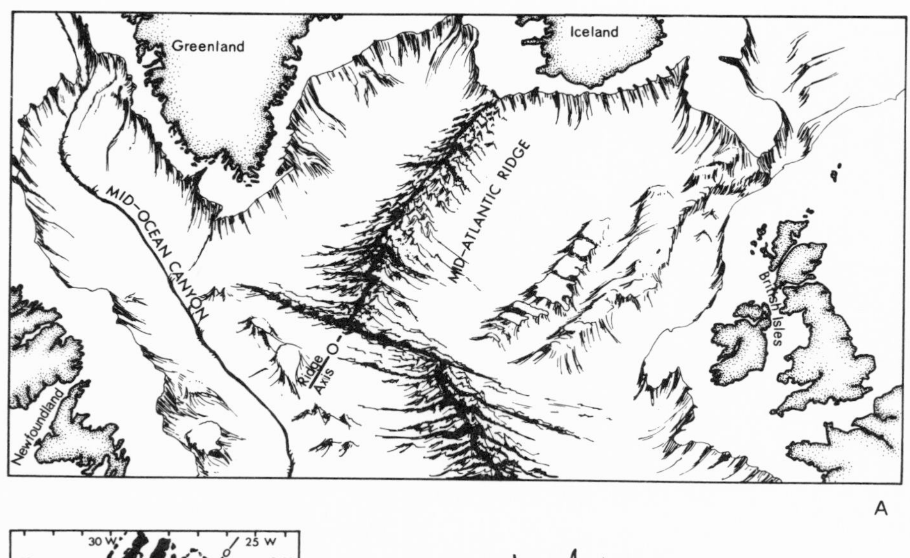

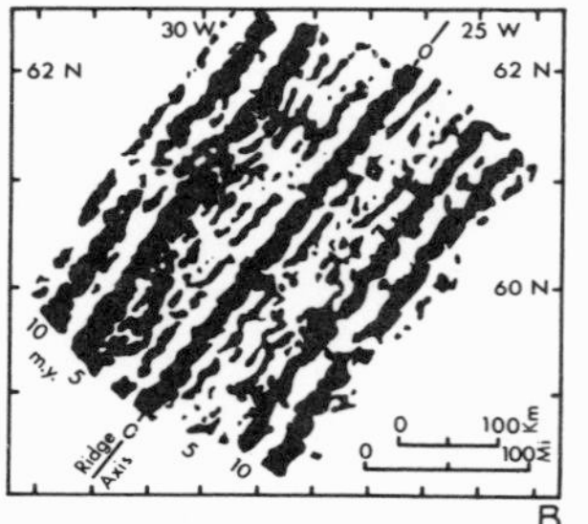

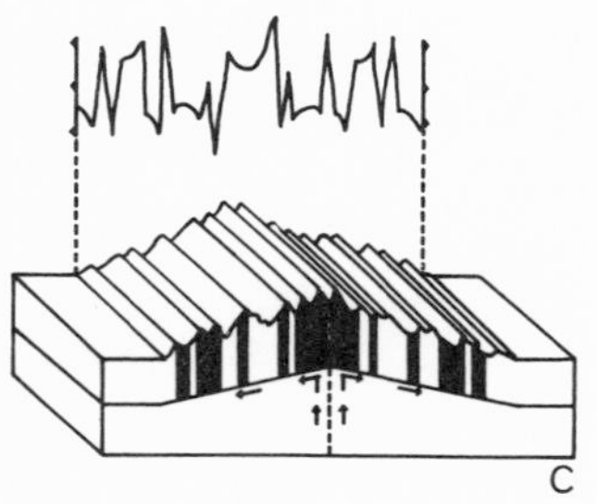

FIGURE 56. A. The North Atlantic Ocean basin, showing the Mid-Atlantic ridge (in this region, known as the Reykjanes ridge), offset from the more southerly part of the ridge by a long transverse fault—the Gibbs Fracture Zone. B. Magnetic anomaly pattern for the Reykjanes ridge. The anomalies form mirror images on the two sides of the ridge axis. C. A. diagram showing the growth of the ridge and the magnetic anomaly symmetry with relation to the ridge axis. The line above the block diagram represents an observed anomaly curve.

cooling. Then, if these rocks were pulled away from the axes of the ridges, they would form bands of frozen magnetism parallel to the ridges. As long as the earth's magnetic field was "normal" this would be so recorded in the rocks that were erupting along the axes of the ridges. But, as the evidence indicates, there would then be a complete reversal of the earth's magnetic field, so that newly erupting rocks, following those that had recorded "normal" magnetism, would show "reversed" magnetism. These rocks, too, would in time migrate away from the axes of the ridges, to form bands of rocks parallel to the ridges. Thus through time there should be a series of magnetic bands, like zebra stripes, paralleling the midoceanic ridges and alternately recording the "normal" and "reversed" magnetic fields of the earth. So much for the theory.

It is possible to measure magnetic anomalies in the ocean floor by towing magnetometers behind survey vessels. As a result of a vast number of measurements in crossing the oceans at right angles to the oceanic ridges, the pattern of magnetic anomalies was revealed. And this pattern showed that, as had been postulated by Vine and Matthews, the measured anomalies did reveal banded zebra-stripe patterns of alternating "normal" and "reversed" fields—these lying parallel to the oceanic ridges. Furthermore, the bands on the two sides of each ridge were mirror images of each other. It was thus quite evident that these bands of magnetic anomalies in the rocks of the ocean floors represented movement through time, away from the ridges, on the two sides of the ridges. What is more, it was evident that these patterns of alternating zebra-stripe ridges are of worldwide extent. (Of course, there are many complicating factors to make the picture of magnetic anomalies far less simple than would appear from this description; but this gives in essence the general aspect of the situation.)

What is the meaning of this worldwide pattern of magnetic anomaly bands paralleling the midoceanic ridges? It means that the ridges actually mark fissures along the ocean floor where molten magmas from the depths of the earth flow upward, like long, vertical plastic sheets. And as they make their way to the surface of the ocean floor the crustal blocks spread to the right and to the left.

By the use of radiometric techniques, it has been possible to measure the time spans of the banded magnetic anomalies. Generally speaking, the "normal" magnetic intervals would seem to last about 420 thousand years, and the "reversed" intervals appear to extend to an average of about 480 thousand years. Measurements have been carried back through time into the Cretaceous period, and some 171 reversals have been identified.

On the basis of these measurements it is possible to estimate the rate of movement of the continental blocks away from each other. However, there are numerous complications. Movement would appear to have been faster in some areas than in others, and at some times as compared with other times through the extent of late Cretaceous and Cenozoic history. All in all, however, the drifting of the continents would seem to have been at the rate of from about half an inch each year, to as much as two inches per year. If, for example, there were to be an average separation of approximately an inch per year, two continental blocks would be separated to a distance of 3,000 miles in about 190 million years. And this is roughly the width of the Atlantic Ocean, and the time span since about the end of the Triassic period. It all goes together.

A corollary to sea-floor spreading is the fact that volcanic islands that arise near the midoceanic ridges are carried away from the ridges by the movement of the crustal plates. Therefore it is not surprising to find that the younger islands are near the ridge, and the older ones are more distant. Indeed, as the volcanoes are carried away from the ridge, down the slope of the ocean floor, they frequently become inactive submarine mounts, called "guyots."

Up to this point our discussion has been concerned with the moving apart of the continental blocks. But what about the opposite sides of these continental blocks? If there is an accretion of material from the depths of the earth, by molten magmas welling up along the axes of the midoceanic ridges and the consequent addition of this material to the oceanic basins—thus causing an increase in the width of these basins—there must be a concomitant loss of and descent of material at some place. For, after all, the earth is essentially a closed system.

The continental blocks as we know them seemingly are only parts of the large, extensive plates, previously mentioned, that form the earth's crust. There are perhaps six major plates: an American plate, made up of the two American continents, and the North and South Atlantic ocean—eastward to the mid-Atlantic ridge, the western edge of this plate being formed by the continental margins where they are met on the north by the Pacific plate, and to the south by the Mid-American trench and the Peru-Chile trench; a Eurasian plate consisting of Eurasia and the North Atlantic Ocean westward to the Mid-Atlantic ridge, and eastward to a series of trenches along the western margins of the Pacific Basin; an African plate, consisting of the African continent, the South Atlantic westward to the Mid-Atlantic Ridge, and including the Indian Ocean eastward to the Mid-Indian Ocean ridge; an Indian plate, made up of peninsular India, Australia and eastward to a series of trenches in the southwest Pacific Ocean; the Pacific plate, consisting of the great Pacific Basin in the southern hemisphere to the west of the long ridge known as the East Pacific Rise—parallel to but distant from the western coast of South America; and the Antarctic plate, made up of the Antarctic continent and the surrounding seas—with northward extending arms, reaching between the

African and Indian plates, and between South America and the East Pacific Rise. Movements and actions are between these plates—as we have already seen in the case of the spreading of the sea floors on each side of the mid-Atlantic ridge.

So it is that these great plates, bounded by and separated from each other in part by the midoceanic ridges, are correlatively bounded by and separated from each other in part by the deep oceanic trenches. Thus the American plate, separated from the Eurasian and African plates on the east by the long mid-Atlantic ridge, is separated in part from the Pacific plate by great oceanic trenches, one of which bounds the western coasts of South and Central America, while another runs along the south side of the Aleutian Islands. Still other trenches range alongside the South Pacific islands, and along the East Indies, while additional trenches bound the western borders of the Pacific Ocean from the East Indies to Kamchatka. There are also trenches through the West Indies and in the South Atlantic. There are probably other trenches that as yet have not been identified. Movement is from the midoceanic ridges toward these trenches.

It would seem probable that the deep oceanic trenches are the regions where oceanic crust is returned to the deeper portions of the earth. To state it in different words, the moving sea floor at the edge of the crustal plate on the side opposite to the midoceanic ridge plunges

FIGURE 57. The pattern of plate tectonics. Six major plates that adequately explain the directions of continental drift are here identified, but additional minor plates are also recognizable. The plates are bounded by ridges or trenches; the arrows indicate directions of movement.

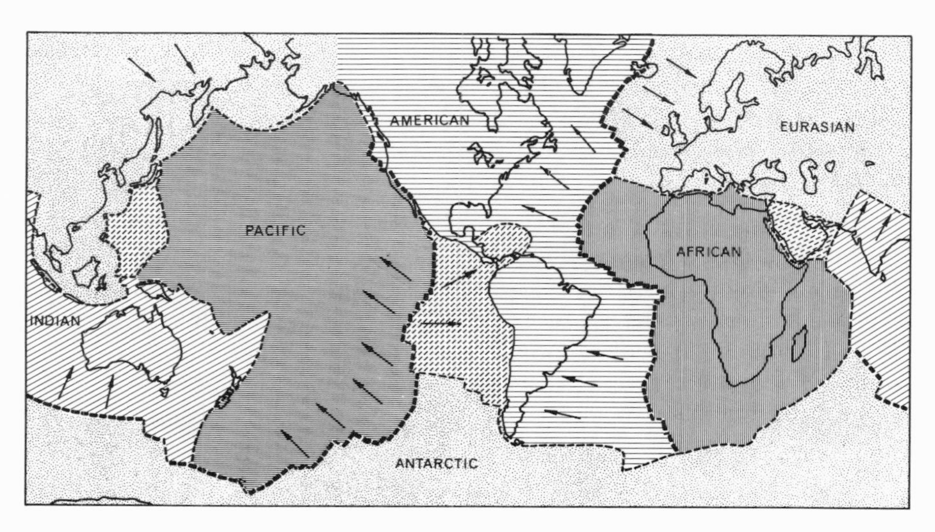

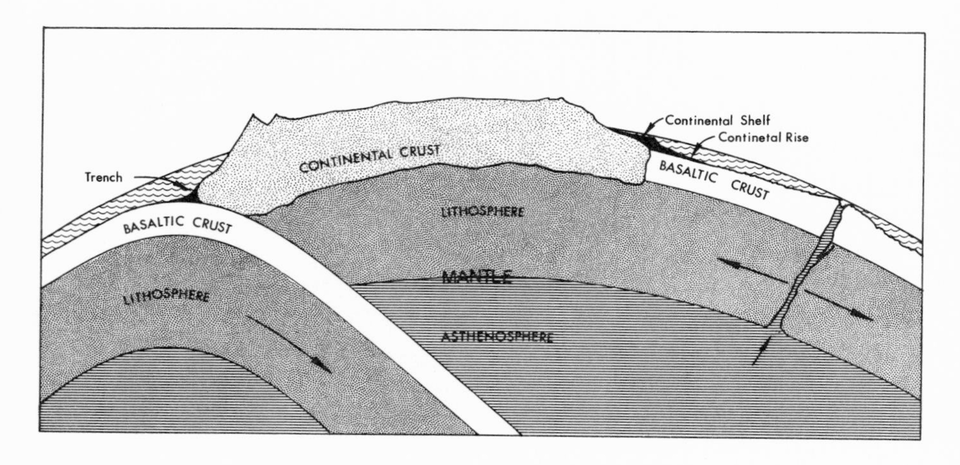

FIGURE 58. A diagram of continental drift. Rifting between two tectonic plates (American and African) results in the welling up of molten rock from the depths, to form the midoceanic ridge (right). The westward drift of the American plate, carrying the continental block, caused a collision with the Pacific plate, this latter being forced down to be destroyed in the mantle. The zone of collision is marked by the pushing up of the Andes and the formation of a deep trench, with resultant deep earthquakes.

down at a relatively steep angle, to underride the plate against which it is being pushed. In this way crustal materials go back into the depths, and in this way the surface area of the earth remains constant.

There are many other factors involved in the movement of the earth plates; for example, the Pacific plate is grinding its way to the northwest in relation to the west edge of North America, causing frequent earthquakes in California. However, these remarks give in essence the probable processes that have caused the separation and the drift of continents during at least 200 million years, since early Mesozoic times. Here, it would seem, is the mechanism that accounts for continental drift. It is a mechanism that basically involves convection currents within the earth—currents that cause deep magmas to rise along the midoceanic ridges, and descend into the depths of the earth as the crust plunges down in the oceanic trenches. It is a mechanism that gives real meaning to continental drift and makes the theory complete.

Plate tectonics, sea-floor spreading, paleomagnetism and paleomagnetic reversals, the fits between the several continental blocks as verified by computerized studies, and the distributions of land-living animals and plants—all of these quite varied lines of evidence are now seen to go together to form the changing patterns of the past, patterns of immense supercontinents that in time fragmented—the fragments, the ancestors of our modern continents, drifting (and frequently rotating as they drifted) to the positions they now occupy across the surface of the globe. With this truly grand concept in mind, let us now turn to some of the details of the disruption of Pangaea, of the long-enduring change that brought about the end of the ancient world and marked the advent of a world directly ancestral to our familiar, modern world.

RIFTING AND DRIFTING OF THE PLATES

As we have seen, there is good reason to think that the rifting of Gondwanaland and Laurasia began when there were great outpourings of basaltic lavas along the edges of what were to become separate continents. The visible evidence of these events is seen, for instance, in the long basaltic ridges, such as the Watchungs of New Jersey and the Connecticut Valley volcanic ridges, that flank the eastern seaboard of North America, or the immense Drakensberg basalts of southern Africa, or the correlative and equally immense volcanic flows in the Transantarctic mountains. These widespread effusions of molten rocks from the depths were the first indications of the profound geological changes which in time were to so greatly alter the face of the earth. These first events took place perhaps about 200 million years ago.

According to one interpretation, recently published by Dietz and Holden (who have been briefly quoted), there were two initial rifts of immense proportions. There was a northern, predominantly east-to-west rift that ran between North and South America and the northwestern bulge of Africa, and between the northeastern coast of Africa and southern Europe, thus separating Laurasia from Gondwanaland, except for a contact between Spain and the Moroccan-Algerian edge of northern Africa. This contact formed a sort of pivot or hinge, upon which Laurasia rotated clockwise to some extent, thereby opening a shallow ocean basin, which eventually was to become the Caribbean Sea and the Gulf of Mexico. The seaway between northeastern Africa and Europe was the long-enduring Tethys Sea, already in existence at the time of the events here being described—a sea which preceded the modern Mediterranean Sea and which, through portions of geologic history, was joined with the Caribbean Sea to form a greatly extended east-to-west ocean basin. Then there was a north-to-south rift, which separated Africa and South America, on the one hand, from the remainder of Gondwanaland, on the other.

At a later date, as some students believe, the block that was to become peninsular India broke away from the Antarctic-Australian segment of Gondwanaland, to begin its long northward drift toward the Asiatic continent. Various authors have envisaged peninsular India as a large, triangular island floating across an incipient Indian Ocean—its base toward the north, paralleling its eventual junction with Asia, a junction that was to be marked by the wrinkling of an edge of the Asiatic land mass to form the ancestral Himalayas. Or perhaps, as has been mentioned, this aspect of the rifting of Gondwanaland involved China as well as India. Whether India alone or India in concert with China drifted to the north, it seems probable that the drifting block was not isolated during Mesozoic time.

There is ample evidence in the form of late Triassic land-living reptiles, to show that there were land connections between peninsular India and the rest of the world; and such connections were present not only during late Triassic time, but also, as is indicated by large dinosaurs, during the Jurassic and Cretaceous periods. Perhaps the separation of peninsular India and possibly China from Gondwanaland was delayed until a much later date. If so, the date necessarily would have been subsequent to Cretaceous history, so that the movement of this land mass across several thousand miles to the Asian mainland would have been remarkably swift.

Whatever may have been the fate of the Indian peninsula and China during later Mesozoic history, we can follow the course of events in other portions of Gondwanaland. It would appear that there was little movement of the African block, a condition that prevailed not only during Jurassic history, but also continued throughout the Cretaceous as well. Africa was, in a sense, a pivot or a base, about which and from which the separation of the other parts of the fragmenting Gondwana continent took place.

It seems likely that the separation between Africa and South America was well under way in the Cretaceous period, so that by the close of the Cretaceous there was a South Atlantic Ocean between Africa and South America, albeit a narrow one, still closed at its northern end.

The opening of the Atlantic Ocean by continental drift affected the North Atlantic as well as the ocean to the south, so that by the close of the Cretaceous there was perhaps a considerable gulf between the Mauretanian edge of the African continent and the eastern seaboard of North America from the Florida region to the area of Nova Scotia.

The Mesozoic North Atlantic Ocean was then continued somewhat to the north, to make a gulf between Spain and the western European coast to the north of Spain, on the east, and the Labrador region on the west. These changes in the relationships of the Laurasian continental elements to the African block would have been brought about by a northwestward drift of North America, coupled with a mild counterclockwise movement or eastward rotation of the Eurasian block, pivoted at the Spanish hinge between Africa and Eurasia. Nevertheless it would seem that Laurasia still remained an entity by virtue of a land connection across what was later to become Greenland, between North America and Europe. This was the broad land bridge that allowed an abundant flow of giant Upper Jurassic dinosaurs between North America and the eastern hemisphere.

While South America was drifting away from Africa, thus opening the South Atlantic Ocean, and simultaneously while North America likewise was becoming separated from its African connection as Eurasia rotated or moved eastward somewhat, to mark the initial opening of the North Atlantic Ocean, the Antarctic-Australian block of Gondwanaland also probably was beginning to drift away from the African core. These now separate continents, still connected, and connected likewise with South America through the Antarctic peninsula, moved as a unit somewhat toward the south and east.

PREMONITIONS OF THE MODERN WORLD

So it was that the rifting of Gondwanaland took place, and the several continental blocks that were formed by the breakup of this ancient southern continent drifted toward the positions that ultimately they were to occupy. Likewise, there was a beginning of the breakup of Laurasia—but only a beginning. Laurasia still remained essentially intact, even though the ancestral North Atlantic Ocean had come into being. But the advent of the modern world had commenced.

Yet the world as we know it, with the continents in the positions so familiar to us, was still many millions of years in the future. The earth would pass through some 60 million years of Cretaceous history, and as many more million years of succeeding Cenozoic events, with the continental blocks continuing to move slowly but inexorably toward their present positions, before the modern spacing of land masses within the surrounding oceans was achieved.

By the end of the Cretaceous period all of the continents as we know them would seem to have been substantially blocked out. The North and South American plate had drifted to the west. The South Atlantic was a comparatively broad ocean, perhaps two-thirds as wide as it is today. On the western border of the South Atlantic was an almost isolated continent—South America—tenuously connected with Antarctica. North America, although beginning to assume the outlines of its present form, was still connected through Greenland to Eurasia. In other words, Greenland was not as yet a distinct and separate land mass. Eurasia, although a single continental mass, was probably superficially divided into eastern and western moieties by a long north-to-south inland sea that traversed the region to the east of the present Caucasus Mountains. Madagascar was by then an island, removed from Africa. The Antarctic-Australian land mass had perhaps withdrawn from close proximity to Africa, with the South Pole located within the confines of East Antarctica. Australia still was connected with East Antarctica, as may have been the case with New Zealand as well.

Such is a summary of the possible chain of events taking place during middle and late Mesozoic times, after the breakup of Pangaea. And if this summary presents a valid picture of changing continental relationships during the long years of the Jurassic and the Cretaceous periods, it outlines a world in transition—from the ancient earth when lands were concentrated in two great connected supercontinents, to the modern earth, where, as we know, the lands form a pattern of scattered but not isolated continents. In the ancient world of Gondwanaland and Laurasia there were broad avenues of communication whereby land-living vertebrates could move from one region to another. In the modern world of six continents—North America, South America, Eurasia, Africa, Australia, and Antarctica—there are restricted avenues of intercontinental communication linking the two Americas in the west and Eurasia and Africa in the east, but with Australia and Antarctica left as isolated land masses.

We have seen how in Permo-Triassic times the presence of the two great supercontinents might have determined the distributions of land-living vertebrates. Moreover, we have seen that this view of ancient animals in an ancient world does appear to coincide with the numerous lines of evidence indicative of the very existence of such an ancient world, so different from the world in which we live. We know, of course, the manner in which the distributions of modern land-living vertebrates are determined by the spatial relationships of the modern continental land masses (subjects which will be accorded some attention later in this book). It is now our problem to look at the continents in transition, the continents of the Jurassic and Cretaceous periods of earth history as they have been interpreted from various lines of geophysical and geological evidence, and to see how such continents accord with the distributions of the land-living vertebrates as we now know them.

This matter of the patterns of life of the land-living vertebrates after the presumed breakup of Gondwanaland and Laurasia, but before the establishment of present-day intercontinental connections, is a most crucial one to the student of animal distributions and relationships. It is a fascinating problem as well, transcending the interests of the specialized scholar. For it deals with the world during the great age of dinosaurs, when giants ruled the earth. And almost everybody is interested in giants.

VI. CONTINENTS AND GIANT DINOSAURS

THE AGE OF GIANTS

The Jurassic period was an age of giants—the first of the two Mesozoic periods (the other being the Cretaceous) when giants dominated the continents and the oceans. The giants on the lands were the great dinosaurs; those in the oceans were the ichthyosaurs, some of which were as large as moderately proportioned whales, the long-necked plesiosaurs, and some marine crocodiles. World environments would seem to have been favorable to giants, and so giants inhabited the world. It is the land-living giants, and the other terrestrial vertebrates that lived with them, that will receive our attention.

Dinosaurs, which had become so widely spread throughout the continents during late Triassic time, continued their worldwide distribution during the Jurassic. They dominated the Jurassic faunas of which they were members, as they were to dominate the faunas of the following Cretaceous period, so that a study of the land-living vertebrates of this portion of Mesozoic history involves primarily the study of dinosaurs and secondarily of those other backboned animals that generally were so overshadowed by the awesome dinosaurs. It must be remembered, however, that the faunas of the Jurassic period were necessarily balanced associations of animals living together, as faunas were before and as they have been since, so that no matter how impressive the dinosaurs may have been in the ecology of Jurassic life we should not allow them to divert us unduly from the other tetrapods that were their neighbors. Yet, in spite of such strictures, it still is difficult to obtain or maintain a balanced view of Jurassic land-living vertebrates, not only because of the large size and the ubiquitous presence of the dinosaurs in the faunas of this period, but also because these great reptiles were fossilized more completely and more frequently than were some of the smaller animals, owing to the nature of the processes of petrification.

This is quite apparent when we attempt to study the land-living vertebrates of early Jurassic age. This phase of earth history would seem to have been a time when lands were topographically low and very probably geographically restricted. Consequently sediments containing the fossilized remains of land-living animals are very sparse—to put it mildly—with the result that our view of the land-living tetrapods living in the early Jurassic world is quite limited and very unbalanced.

Such glimpses as we are able to obtain show us a world in which those "new" animals that had arisen during Triassic time continued to prosper. As we have seen, the close of the Triassic period saw the extinction of various "old" amphibians and reptiles, the descendants of long-established lines of Paleozoic ancestors, that had managed to hold on during Triassic times. Such were the labyrinthodont amphibians, the last of those primitive reptiles known as cotylosaurs, embodied in the small lizardlike procolophonids, the rather enigmatic and generally infrequent protorosaurs, and, of course, almost all of the mammallike reptiles. The words "almost all" are used in this connection because two restricted groups of mammallike reptiles, the highly specialized and rather rodentlike tritylodonts and the very small and very advanced ictidosaurs (which were in certain respects connecting links between reptiles and mammals), held over into the beginning of Jurassic history. Finally they, too, succumbed, perhaps as ecological victims of the early mammals, to which they were so closely related.

So it was that the early Jurassic continents were inhabited, in addition to the doomed tritylodonts and ictidosaurs, by several groups of dinosaurs, early turtles, crocodilians, and the first flying (as opposed to gliding) backboned animals. There must have been rhynchocephalians and lizards as well, since their fossils are known from Upper Triassic sediments, while similarly there must have been frogs to represent the amphibians. There were also, it should be added, very primitive mammals.

The dinosaurs, which had their true beginnings with the advent of late Triassic history, were taxonomically rather restricted during the final years of the Triassic, even though they were geographically widely distributed. In other words, the dinosaurs of the Triassic were comparatively primitive (for dinosaurs) and were limited in the range of their adaptations for different modes of life. Some of these first dinosaurs were quite small, as might be expected; others were of considerable size, thus showing the trend toward giantism that was so characteristic of dinosaurian evolution.

Since much of the discussion in this chapter and the next will be concerned with dinosaurs, perhaps it may be well to devote a few lines to these interesting reptiles, in order that the reader who is not familiar with them may have a reasonably clear idea as to their constitution and relationships. The dinosaurs are renowned as Mesozoic giants, and a majority of them were giants; but some of them were small. This points up the fact that the dinosaurs cannot be defined by their size (except incidentally) but rather must be understood upon the basis of their anatomical structure. For example, one does not define a dog because of its size—for there are very large and very small dogs—but by virtue of those certain characteristics that we (as well as the dogs themselves) recognize as definite doggy traits. The point need not be labored.

The dinosaurs are not a single group of reptiles, but are composed of two distinct orders, the *Saurischia* and the *Ornithischia*, separated by many anatomical features, among which the structure of the pelvis is of particular importance. In the *Saurischia* the pubic bone is in the form of a forwardly extending rod or plate; in the *Ornithischia* this bone is a relatively thin and often reduced rod, rotated backward to lie contiguous and parallel to the backward-extended ischium. Both of these dinosaurian orders appeared in the middle or later portions of Triassic history, as descendants from earlier Triassic thecodont reptiles. As they evolved, from late Triassic time through the Cretaceous period, the saurischian dinosaurs followed a tripartite pattern of development.

Among the first of the saurischians to appear in the Triassic record were the primitive theropods, which to begin with were small, delicately built dinosaurs that walked on birdlike hind limbs and used the small forelimbs and the flexible hands for the purposes of grasping—possibly to a considerable degree as aids in the gathering of food. The skull and jaws were armed with sharp teeth. Obviously these little dinosaurs were carnivorous. Some of these small theropods continued throughout Mesozoic time as the hunters of small game; these are known as the "coelurosaurs." But from coelurosaurian ancestors there evolved the giant carnivores, the predators of other giant dinosaurs, the carnosaurs of Jurassic and Cretaceous times. They retained in essence the adaptations of their small ancestors, but, of course, on a magnified scale. These giant predators were characterized especially by their huge skulls, the long jaws of which were set with large, saberlike teeth, and in many cases by their ridiculously small forelimbs. The theropods were the only carnivorous dinosaurs; all other dinosaurs were herbivorous.

Also living in the Triassic were those saurischian dinosaurs known as "prosauropods." These dinosaurs quickly evolved in the direction of giantism, and some of them were as much as 20 feet or more in length. They were heavily built, yet even so they were dominantly bipedal, walking on the hind limbs. The skull was relatively small, and the teeth were bladelike. Evidently they fed upon plants.

The prosauropods would seem to have been ancestral to the gigantic sauropod dinosaurs of Jurassic and Cretaceous age. These, the largest animals ever to have lived on the land, were giants among giants, often attaining lengths of 60 or 70 feet and live weights of 40 or 50 tons, and even in one case, perhaps of some 80 tons or more. Being so large they necessarily were quadrupedal, using all four limbs, which were very strong, for support. The neck and the tail were elongated, and the comparatively small head was supplied with weak jaws and small teeth. It would appear that these dinosaurs frequented swamps and fed upon soft vegetation.

Whereas the saurischian dinosaurs evolved along three principal lines of development, namely, the theropods, prosauropods, and sauropods (each, of course, containing various lesser radiating lines), the ornithischian dinosaurs followed four major evolutionary lines. These were the ornithopods, or duck-billed dinosaurs; the stegosaurs, or plated dinosaurs; the ankylosaurs, or armored dinosaurs; and the ceratopsians, or horned dinosaurs. All of these dinosaurs evidently were herbivores from the beginning of their several histories. The first ornithischians were very small ornithopods, known from the Triassic of South Africa. Although these little dinosaurs were as yet but slightly advanced along the path of ornithischian evolution, they possessed various characteristic ornithischian features, among which may be noted the structure of the pelvis and the possession of rather bladelike teeth in the sides of the jaws, adapted for the cutting of vegetation. This

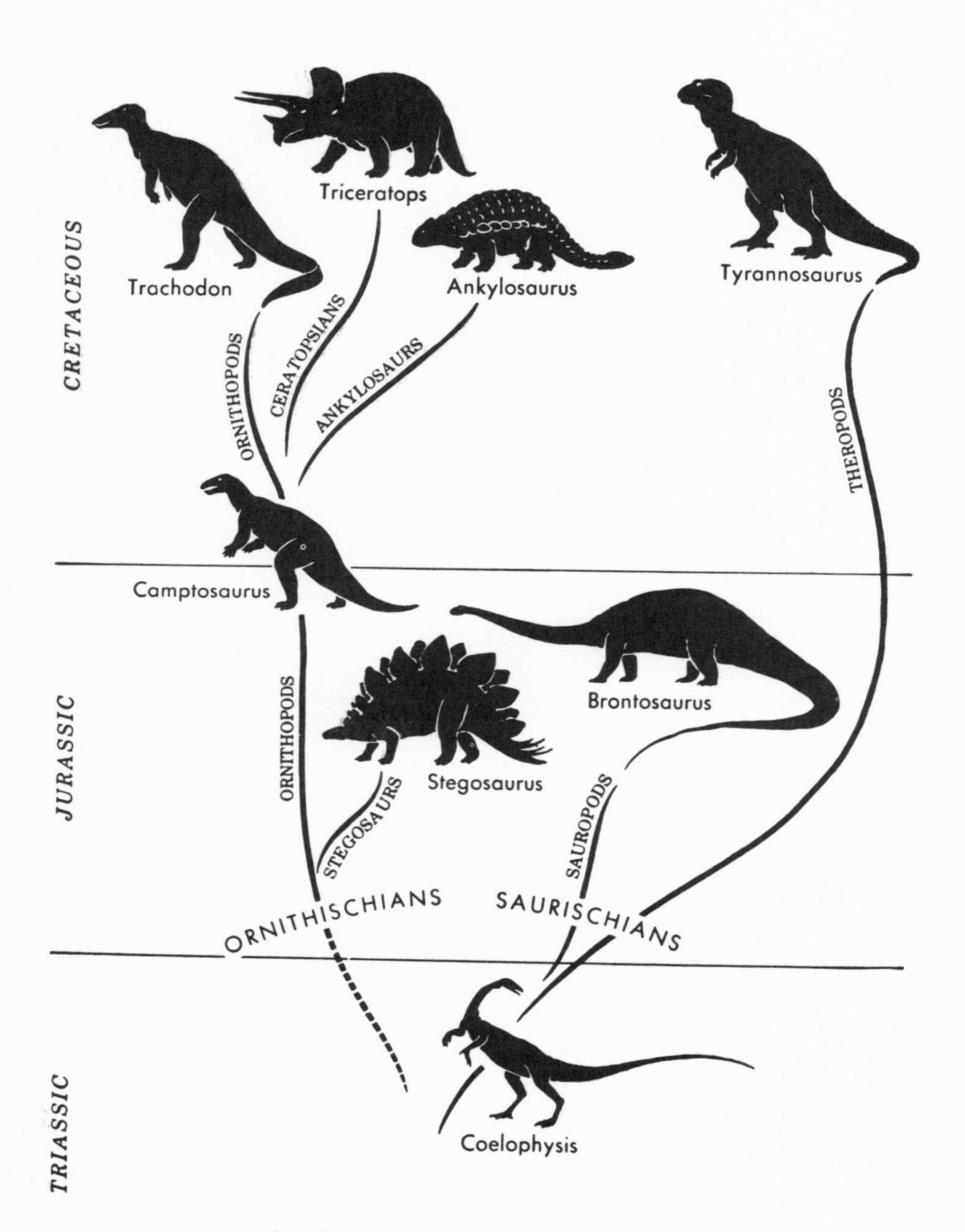

FIGURE 59. The evolutionary relationships of the dinosaurs.

last is important in dinosaurian ecology. Plant eating among certain saurischians was imposed at a subsequent date over ancestral carnivorous adaptations; among the ornithischians it was a basic adaptation.

The dominantly bipedal ornithopods were the well-known duck-billed dinosaurs and their relatives, many of them adapted for life in the waters of rivers and lakes, where presumably they may have fed in part upon aquatic vegetation. (There is reason to think that they also browsed upon coniferous trees.) The stegosaurs, essentially limited to the Jurassic period, although some of them did straggle across the time line into the opening phases of Cretaceous history, were obviously dwellers on firm ground, where they browsed upon the plants of the uplands. These dinosaurs were protected after a fashion by horny scutes on the body, and they frequently had large, vertical flat plates down the middle of the back and long, sharp spikes on the end of the tail. Such was the equipment of *Stegosaurus*. The armored dinosaurs were, as their name suggests, heavily protected by bony armor (in life obviously covered by horny scutes) that covered the head and a large part of the body. Many of these dinosaurs were equipped with massive, cruel clubs at the end of their tails, by means of which they could flail enemies that might attack them, with devastating effect. The horned dinosaurs were in a way the "rhinoceroses" of the Mesozoic. These massive dinosaurs, quadrupedal in pose, as were also the stegosaurs and ankylosaurs, were characterized by huge frilled skulls, on which commonly there were long horns. Thus the ceratopsians could drive off their attackers by vicious lunges. All of these ornithischians, being herbivores, were essentially harmless reptiles, but in a world inhabited by giant carnivorous dinosaurs they were variously equipped to flee or to protect themselves.

Such were the reptiles that ruled the continents of the Jurassic and Cretaceous world. With this introduction to the dinosaurs, we may now return to the problem of Jurassic continents and the distributions of land-living tetrapods on those continents.

DINOSAURS AND OTHER REPTILES OF EARLY JURASSIC TIME

As has been mentioned, the beginning of the Jurassic was marked by low and restricted lands. The continental blocks with which we have been concerned were still present, as they had been in the Triassic, and presumably were drifting in the fashion and in the directions that have been described in the preceding chapter. But even though the con-

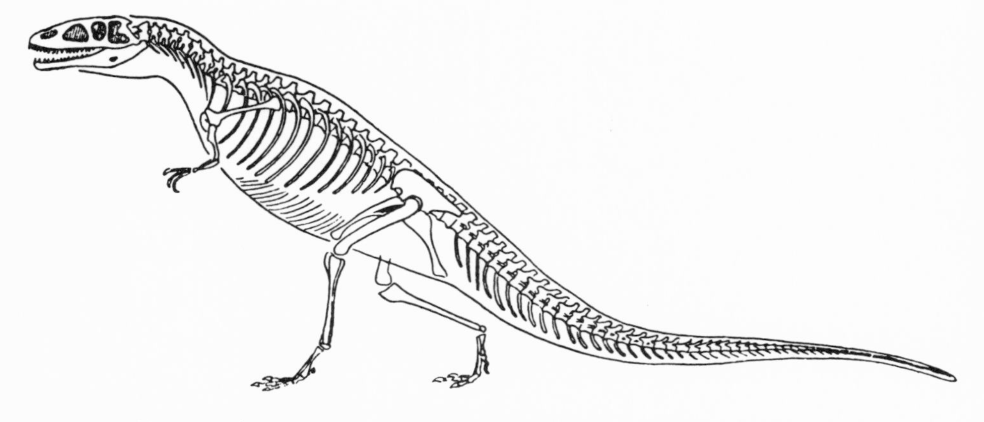

FIGURE 60. *Megalosaurus,* a Lower Jurassic carnivorous dinosaur from Europe.

tinents were basically present as large blocks rifting from the parent Gondwanaland and Laurasian land masses, many of them were superficially reduced in area by shallow marine flooding of their surfaces. This was particularly true in Europe, which at the beginning of Jurassic history would seem to have been a region of shallow tropical seas, forming numerous embayments within the continental blocks. There were many peninsulas, archipelagos, and isolated islands, as one might expect. This is why so many marine forms are found in the Lower Jurassic deposits of Europe—varied seashells, fishes, ichthyosaurs, and plesiosaurs. This is why the record of land life is so restricted.

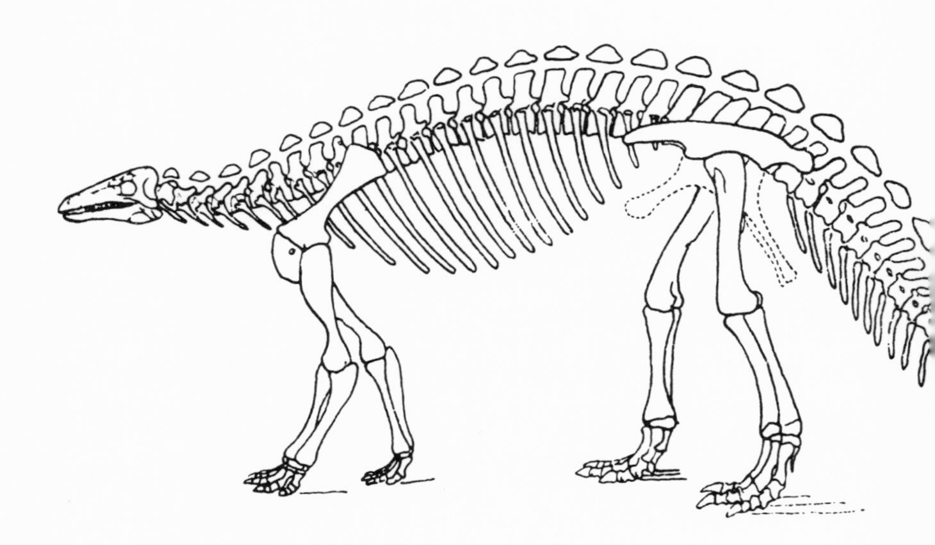

But here and there some land-living reptiles have been found, especially in the Lower Jurassic, or Lias, of England and northern Europe. Two dinosaurs from the Lias of England afford a fragmentary clue as to the nature of these Mesozoic rulers in this part of the world at the beginning of Jurassic history. One is *Megalosaurus,* a gigantic carnivorous predaceous carnosaur; the other is *Scelidosaurus,* the earliest known stegosaurian, already well armored with protective scutes on the body, as a defense against its megalosaurian contemporaries. *Megalosaurus* was a bipedal reptile, as were the other carnivorous dinosaurs, walking on hind limbs that were curiously birdlike, in spite of their large size and massive construction. The forelimbs were very small; the head was very large, thus affording a wide gape for the jaws, carrying their generous complements of cruel teeth. The long tail made a counterweight for the mass of the body in front of the pelvic articulation. *Scelidosaurus,* although of medium size in dinosaurian terms, was quadrupedal; the weight of its armor plate surely dictated such a pose. The head was small as were the serrated teeth, adapted for cutting vegetation.

It seems obvious that there were many more dinosaurs living in the Liassic faunas of northern Europe, but our record is a most incomplete one. Certainly there were turtles at this time on the northern continents, while the edges of rivers and lakes, as well as the extensive tropical seashores, were inhabited by early crocodilians. We can infer

FIGURE 61. *Scelidosaurus,* a Lower Jurassic armored dinosaur from Europe. This dinosaur and *Megalosaurus* (shown in the preceding figure) afford some clues as to the directions of dinosaurian evolution that developed during Jurassic and Cretaceous history. It was an evolutionary pattern typified by growth to giantism; giant plant-eating dinosaurs filled Jurassic and Cretaceous landscapes, and they were hunted by giant predators.

FIGURE 62. The classic Lower Jurassic, or Liassic, beds of the English Channel coast. From these banded near-shore deposits, in the vicinity of Charmouth, have been found the skeletons of marine ichthyosaurs and plesiosaurs, of flying pterosaurs, and an occasional land-living dinosaur, such as *Scelidosaurus*.

also that there probably were numerous lizards, and among the amphibians, frogs. Yet again the record is incomplete.

In the Liassic of England and of central Europe there have been found a few skeletons of flying reptiles, or pterosaurs, evidently the remains of animals that fell or were washed into the quiet waters of shallow lagoons, to be covered by mud and thus fossilized. These early Jurassic pterosaurs were of small size—no larger than sparrows or crows. In these lightly constructed reptiles the fourth finger of the hand was greatly elongated to serve as the support for a leathery wing membrane. Evidently they were able flyers, admittedly not restricted to the land surfaces, for there is ample evidence that they flew across seaways as well as over the land. But perhaps it is valid to include them in our consideration of land-living vertebrates, since they certainly were not confined to the oceans.

From such composite and incomplete sources, and from well-justified inferences, we can thus obtain a view of early Jurassic life in what is now northern Europe. It seems evident that large dinosaurs dominated the scene, as they would for the next hundred million years

FIGURE 63. Giant dinosaurs were living in India at the beginning of Jurassic history. Here Dr. S. L. Jain and Mr. Pronob Mazumdar of the Indian Statistical Institute are excavating a large limb bone from the Lower Jurassic Kota Formation, near Sironcha, in the Godavari-Pranhita Valley.

of earth history. And the tetrapods that shared the lands with them were various crocodilians and turtles, as well as early lizards and frogs. Small and very primitive mammals also inhabited the undergrowth or sought the protection of rocks and trees, well out of reach of the dominating reptiles. Curious aerial reptiles, the pterosaurs, flew through the air and probably lived along the cliffs that overlooked the widely expanded Jurassic seas.

An interesting discovery of land-living reptiles of early Jurassic age has in recent years been unearthed in peninsular India. The Kota Formation there, a rock unit of very early Jurassic affinities, has yielded abundant bones of large sauropod dinosaurs. The fossils have not as yet been scientifically described, but even at first glance they show us that some dinosaurs had become giants by the beginning of the Jurassic.

The presence of these gigantic dinosaurs in the Kota beds is interesting not only because it demonstrates the rapid attainment of gigantic size within this line of reptilian evolution, but also because it shows that India still maintained connections with other parts of the world during early Jurassic time. Certainly the Kota dinosaurs were not isolated; one cannot imagine them as having originated independently of other dinosaurs. They obviously had lines of communication to other lands.

Beyond these discoveries, there is little in the way of Lower Jurassic records to show us what land-living tetrapods were like in other parts of the world. In the Navajo sandstone of Arizona, a formation of lowest Jurassic (and perhaps of uppermost Triassic) affinities, there has been discovered the partial skeleton of a small theropod dinosaur. The Navajo sandstone is a dune sand, indicative of a desert environment. Here we have a most fleeting glimpse—and that is all—of an early Jurassic environment and its tetrapod life in North America.

An equally fleeting glimpse of possible early Jurassic life in Asia is to be had from some dinosaurian footprints, designated as *Jeholosauripus,* discovered in northern China. These tracks represent a small theropod, probably a coelurosaurian, and although they may be of early Jurassic age, they also may come from Upper Triassic sediments. Some carnosaurian footprints are known from Lower Jurassic deposits in Morocco. The one other record that may be mentioned here is of a sauropod dinosaur, *Rhoetosaurus,* known from bones found in Queensland, Australia.

Beyond this there is little to be said. The records of early Jurassic land-living tetrapods in Asia, Africa, and Australia are indeed fragmentary, as we have seen is the case in North America and likewise in South America, being restricted on this last continent to an early frog.

This review of Lower Jurassic land-living tetrapods obviously is a rather dismal account (from the paleontologist's point of view) of limited localities yielding sparse fossils. Yet there is every reason to think that land-living amphibians and reptiles inhabited the continents probably in abundance at this stage of earth history. Here we encounter one of the hazards of paleontology—the sometimes inadequate preservation of fossils. That this is especially marked as it pertains to Lower Jurassic land-living tetrapods probably is owing in a considerable degree to the spread of shallow-water seas over large continental areas at the opening of Jurassic time.

LOWER JURASSIC TETRAPODS AND CONTINENTS

How does the distribution of Lower Jurassic amphibians and reptiles, as we know them, fit with the relationships of the continents that has been outlined? Simply stated, the correlation would seem to be good. There were avenues for the overland movements of amphibians and reptiles between northern Europe and North America, and northern Europe, China, and northern Africa, and it would seem possible from Africa into Australia through the still-connected Antarctic-Australian continental block. As yet no land-living tetrapods of Jurassic age have been found on the Antarctic continent, but the presence of undisputed Lower Jurassic fresh-water fishes in the Transantarctic mountains indicates that conditions may have been right for the presence of tetrapods in this region. Perhaps *Rhoetosaurus*, the Lower Jurassic sauropod found in Queensland, reached the region of its interment by an African-Antarctic route.

Or can its presence in Australia be correlated with the presence of sauropod dinosaurs in the Kota beds of peninsular India? Could not India still have been joined with Africa on the one side, and with Antarctica and Australia on the other? In this connection it is interesting to note that in speaking of late Jurassic relationships (which might very well have held in the early Jurassic as well) Furon remarks that "India can have been in connection with Madagascar" (Furon, 1963,

FIGURE 64. A reconstruction of Pangaea during Jurassic times. The ligation of India and Antarctica-Australia to Africa is shown as persisting, in light of the known distributions of Jurassic dinosaurs.

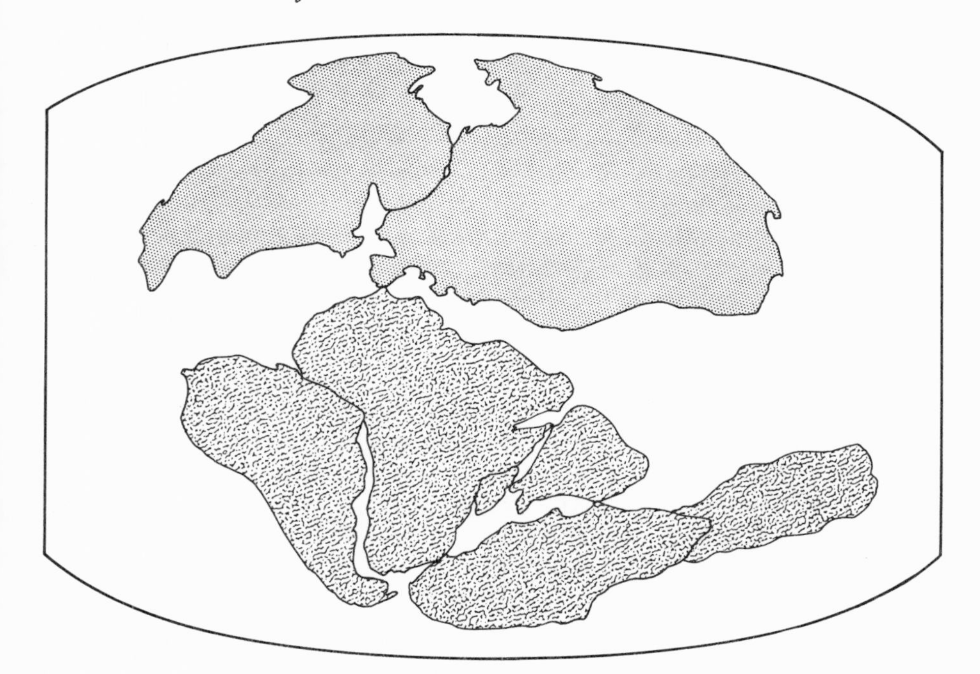

p. 361). If such were the case, then the presence of Lower Jurassic sauropod dinosaurs in India and in Queensland is readily explained.

THE MIDDLE JURASSIC RECORD

Our knowledge of Middle Jurassic land-living vertebrates, and therefore of the distribution of these animals on the continents of that time, is perhaps more restricted than it is for Lower Jurassic tetrapods. The fossil record does leave much to be desired.

However, the continued domination of the continents by large dinosaurs is shown by the presence in France of the widely dispersed carnosaur *Megalosaurus* and in Morocco of large carnivorous theropods and of gigantic sauropods. Far to the southeast, in Madagascar, the bones of large sauropod dinosaurs occur in sediments of this age. Again, far to the east, in China, footprints of coelurosaurian and carnosaurian dinosaurs are contained in rocks attributed to the Middle Jurassic. In southern Argentina there are some footprints of Middle Jurassic age that have been attributed to small theropod dinosaurs. In addition, the bones of a sauropod dinosaur—one of the gigantic swamp-dwelling dinosaurs that were so characteristic of Jurassic history—have been found in Chubut Province of southern Argentina. This dinosaur, named *Amygdalodon,* is a rather primitive type of sauropod, and the suggestion has been made that it may be related to *Rhoetosaurus,* the Lower Jurassic dinosaur found in Australia. In addition to these dinosaurs, a Middle Jurassic frog has been found in Argentina. Thus a record of land-living vertebrates in the Jurassic of South America does exist, but as is the case of Lower and Middle Jurassic amphibians and reptiles in other parts of the world, it is a scant representation indeed.

In the Stonesfield slates of southern England, fine-grained deposits that seemingly accumulated in an estuarine environment near the shore, are the tiny and delicate jaws, teeth, and occasional bones of Middle Jurassic mammals. These fossils are extraordinarily important, if not spectacular, for they give us a glimpse of the primitive mammals of middle Mesozoic times.

Such is the evidence of Middle Jurassic land-living vertebrates as it is now known, and as such it demonstrates that life on the land was developing toward its spectacular manifestations of late Jurassic time. Moreover it demonstrates that, so far as we can see, the avenues for the distribution of land-living tetrapods between the rifting fragments of ancient Gondwanaland and Laurasia remained in being. Giant dino-

FIGURE 65. Excavating giant dinosaur bones from the Upper Jurassic Morrison Formation in Utah. The scene is at Dinosaur National Monument some years ago, before the handsome building, now familiar to thousands of visitors, was built over the quarry. The paleontologist is Gilbert Stucker, now of the American Museum of Natural History.

saurs in China, Europe, northern Africa, Madagascar, and South America certainly are evidence of this.

DINOSAURS OF THE LATE JURASSIC WORLD

It is no exaggeration to say that the fossil skeletons of many Upper Jurassic land-living vertebrates are indeed spectacular. For in certain deposits from this section of the geologic column are found the massive bones of giant dinosaurs in abundance and considerable variety. The remains of Upper Cretaceous dinosaurs are perhaps the most varied of all dinosaurian fossils, as we shall see, but for overwhelming size that can only impress and awe the human mind, the petrified bones of some of the Upper Jurassic dinosaurs are without equal in the world of paleontology.

These dinosaurs, which, it should be repeated, constitute only a part, but certainly the dominating part, of the faunas to which they belong, are found in four principal regions of occurrence. In the first place there are the dinosaurs of the Morrison fauna, contained within the Morrison Formation of western North America. The Morrison Formation is widely exposed, but the principal collecting grounds have been at what is now the Dinosaur National Park in Utah and Colorado, at the famous Como Bluff locality in Wyoming, at the equally famous Bone Cabin quarry site and adjacent localities, also in Wyoming, at the Cleveland Quarry in the San Rafael Swell of Utah, and at localities near Canyon City, Colorado. Morrison vertebrates have been found at various other localities, but these particular ones have yielded especially large collections.

Another rich collecting ground, and one far removed from western North America, is in Tanzania, in eastern Africa, where the renowned Tendaguru deposits were excavated on a large scale, to yield spectacular skeletons of gigantic dinosaurs, during the first decade of the present century. Again, Upper Jurassic vertebrates are rather well represented at various localities in Europe from Portugal to southern England. This region has not provided rich quarry deposits, such as those of the Morrison and the Tendaguru beds, but in sum the Upper Jurassic of western Europe has yielded a considerable quantity of terrestrial fossils of this age. Finally, Upper Jurassic land-living tetrapods have been found in some abundance in China during the past few decades, especially in the province of Szechwan. Thus it is that from these four regions, North America, Africa, Europe, and Asia, there have been amassed large collections of fossils and from them a wealth of information concerning the backboned animals that inhabited various Jurassic continents.

There are still some lacunae. Nothing is known as yet about the Upper Jurassic land-living tetrapods of South America, and the same is true for Antarctica. The information from Australia is of the scantiest nature. Yet perhaps what we have from the Morrison and Tendaguru deposits, from Europe and from China, in a sense makes up for our lack of knowledge concerning the other continental regions.

The rather abundant evidence about tetrapods on Upper Jurassic continents, as contrasted with the dearth of information concerning these vertebrates in Lower and Middle Jurrasic sediments, is perhaps owing in large part to the favorable conditions for the interment and preservation of animal remains during the latter years of Jurassic history. It seems probable that, although lands were generally close to sea level in the later years of the Jurassic, they were still sufficiently elevated so that incursions of shallow seas across the continents were much restricted as compared with what had been the case in early and middle Jurassic time. Moreover, the world was largely a tropical world, which favored reptilian giants. The land surfaces were widely covered with tropical and subtropical forests, cut by numerous broad rivers and interspersed with numerous shallow lakes, which again favored reptilian giants. The rivers and the lakes frequently were the locales for rapid accumulation of sands and muds, and such conditions favored the frequent burial and preservation of plant and animal remains. Consequently the Upper Jurassic deposits of the regions that have been cited are rich hunting grounds for the paleontologist.

An interesting feature of the Morrison fauna is the large number of gigantic sauropod dinosaurs it contains. Not only are specimens abundant, but also there is a considerable variety of genera and species of these impressive reptiles, including such typical forms as *Apatosaurus, Barosaurus, Brachiosaurus,* and *Diplodocus*. These dinosaurs were all of a similar pattern; they were anatomical variations on a single theme. In all the neck and tail were long, the body was heavy, and the limbs were very massive to support the weight of these huge beasts. In all the skull was relatively small and the teeth were weak—obviously adapted only for soft vegetation. In all the nostrils were elevated to the top of the skull, as an adaptation for breathing while the body was largely submerged. *Diplodocus* differed from the other forms in being a very slender—one might say an attenuated—sauropod. In contrast, *Brachiosaurus* was the giant of them all, the largest of any land-living animal that has ever lived. It was noteworthy by reason of its large forelimbs, causing the shoulders to be much higher than the pelvic region, so that the back was sloping. Of course, where there were such numbers of giant herbivores, there were carnivorous giants that preyed upon them, and so the Morrison fauna contains its complement of large, predatory carnosaurs, such as *Allosaurus* and *Ceratosaurus*. These giant meat eaters were in many respects similar to each other. Both were some 30 feet or so in length; both walked on strong hind limbs; both had long, tapering tails; both had small forelimbs, with hooklike claws that could aid in catching prey and feeding; both had enormous skulls armed with sharp teeth. *Ceratosaurus* was distinguished by a horn on the nose. There were also small coelurosaurs as well, of which the genus *Ornitholestes* is an example, that must have preyed upon small reptiles and other animals of the undergrowth. *Ornitholestes,* no more than five or six feet in length, retained many of the features of its primitive dinosaurian ancestors. It was a light, active, very birdlike dinosaur in structure. The skull was small and the teeth were sharp. The forelimbs were relatively large, although not effectual for locomotion, and the long, clawed fingers would have been very efficient in grasping small prey. The Morrison fauna also contains a few ornithischian dinosaurs, primitive ornithopods in the form of the genus *Camptosaurus*, a relatively small bipedal ornithischian dinosaur, in which the jaws of the small skull were set with primitive plant-cutting teeth, and the well-known plated dinosaur *Stegosaurus*, of which a description already has been made. These are the animals that dominate the fauna.

There is, however, much more variety in the Morrison fauna than is indicated by the dinosaurs, varied though they may be. There are crocodilians and turtles, which if dinosaurs were not present would be regarded as the giants of the assemblage. Lizards are present as well, and rhynchocephalians. One small bone fragment, slight evidence in itself, is nevertheless sufficiently diagnostic to show that flying reptiles, or pterosaurs, were living in the Morrison landscape. (This is corroborated by the discovery of a rather good pterosaur skeleton from Upper Jurassic deposits in Cuba, and, of course, pterosaurs are well known from the Upper Jurassic of Europe.) The amphibians of the Morrison fauna are represented by frogs. Finally, at one site along Como Bluff, Wyoming, known as "Quarry Nine," there have been found the fossils of various early mammals. These fossils, quite small and delicate, and preserved by virtue of the nature of the sediments deposited at the Quarry Nine locality, are minuscule as contrasted with the massive dinosaur bones found nearby, but their significance is great. Here are the remains of animals representing the Class of vertebrates that millions of years in the future was to inherit the earth, when the dinosaurs became extinct.

Although there undoubtedly were other animals, in addition to the ones mentioned comprising the Morrison fauna, the fossils so far recovered are sufficiently comprehensive to give a reasonably good impression of an ecologically balanced fauna typical of late Jurassic history.

That this is so is demonstrated by the fossils collected in Europe, in China, and at Tendaguru in Africa. In all of these localities there are found large sauropod dinosaurs generally similar to and in some cases identical with those of the Morrison deposits. It is especially interesting and informative to note, for example, that *Brachiosaurus*, the greatest of the sauropods and the largest animal ever to live upon the land, occurs in the Morrison beds, in Europe, and in the Tendaguru deposits.

FIGURE 66. A late Jurassic scene in North America, as based upon knowledge of fossils from the Morrison Formation. In the left foreground is the plated dinosaur *Stegosaurus*, looking toward the giant predator *Allosaurus*, on the right. Between them is the small theropod dinosaur *Ornitholestes*. In the background are two giant sauropod dinosaurs, *Brontosaurus*, and beyond them more sauropods, *Diplodocus*. Turtles lived with the dinosaurs of Jurassic times.

FIGURE 67. One of the quarries from which quantities of gigantic Jurassic dinosaur bones were excavated many years ago in what is now Tanzania, Africa. These dinosaurs, closely related to the Upper Jurassic Morrison dinosaurs of North America, indicate the scope of dinosaurian distributions across Pangaea.

The same generally is true for the other dinosaurs, the carnivorous forms, the camptosaurs and the stegosaurs.

Crocodilians are found in Europe, Asia, and Africa. Lizards and rhynchocephalians occur in Europe and China; their absence from the Tendaguru fauna is probably an accident of preservation, or of collecting. Pterosaurs, which occur in the Tendaguru deposits, are well known from the Upper Jurassic of Europe, being beautifully preserved in the fine-grained limestones of Solnhofen, Bavaria. These sediments, and similar deposits elsewhere in Germany and France, were laid down in the still waters of a marine lagoon, where flying reptiles occasionally dropped from the skies into the sea, to be drowned and buried. So fine are the sediments in which these pterosaurs occur that impressions of the wing membranes are preserved.

Three fossil treasures of rare value are the skeletons, with impressions of the feathers, of the first bird, *Archaeopteryx,* preserved in these same Solnhofen limestones. Early birds have not been found in other sediments of late Jurassic age, but this is almost surely an accident of fossilization. At best, birds are rare in fossiliferous deposits, and one would hardly expect these first birds to have been preserved except under the most favorable conditions.

Finally, there should be mentioned the fact that the Upper Jurassic of Europe has yielded turtles, frogs, and primitive mammals, to match the occurrences of these groups of animals in the Morrison deposits.

WORLDWIDE MIGRATIONS OF LATE JURASSIC REPTILES

From this review it can be seen that there are remarkable similarities between the Upper Jurassic faunal assemblages of North America, Europe, China, and eastern Africa. In a few cases identical genera are found at these far-flung localities; in many cases the genera that comprise the several faunas are closely related to each other. In all cases, making allowances for certain absences due to the lack of preservation of specimens, it would seem that the faunas are similar. From them one has a good record of late Jurassic tetrapod life on the continents of that distant day. To recapitulate, this life consisted of gigantic sauropod dinosaurs, giant carnosaurs as well as small coelurosaurs, primitive ornithopod dinosaurs and plated stegosaurs, varied crocodilians, flying pterosaurs, lizards and rhynchocephalians, turtles, frogs, very primitive birds, and likewise very primitive mammals.

That these tetrapods should be so widely dispersed points to two conclusions. One is that continental connections were such that broad avenues for dispersal allowed the movements of full faunas from one region to another. There is no apparent filtering in the expression of these faunas, as one would expect if the connections between land masses were in the form of narrow isthmian links. The second conclusion is that the late Jurassic world was certainly a largely tropical world. It seems probable that the huge reptiles, so dominant in the Upper Jurassic faunas, could have lived only in benign climates. Geographically, climatologically, and biologically it was one world, in spite of the constant rifting and drifting that must have been taking place.

The evidence would seem to indicate that the connection between North America and northern Europe, through what later would become Greenland, remained broad and firm, as it would continue to be for many millions of years into the future. The North Atlantic was narrow and was still limited to its southern reaches. Moreover, the ancient connection between Gondwanaland and Laurasia, through the Spanish-Moroccan hinge, still held. Furthermore, Madagascar probably had not as yet broken away from the African mainland, and, as was pointed out in a preceding paragraph, there may very well have been a connection between Madagascar and peninsular India. Finally, it is possible that what is now the eastern border of peninsular India was still joined with Australia.

A world thus composed would allow ample scope for the wide migrations of terrestrial faunas. Dinosaurs and the other tetrapods that lived with them could have spread their distributions back and forth between North America and Europe, between Eurasia and Africa, and probably through India on into Australia, where one dinosaur, *Agrosaurus,* a small coelurosaur, is known from Upper Jurassic beds. It is quite probable that Antarctica was still joined in some degree to Africa and was involved in these broad movements of land-living tetrapods, but the evidence for this has yet to be discovered. The same is true for South America, which, it would seem, was still joined across its northern Brazilian portion to Africa. The South Atlantic was opening, from south to north, but still would seem to have been very narrow.

This view of late Jurassic continents is fully reconcilable with our knowledge of the distribution of late Jurassic land-living vertebrates (the only stage in the Jurassic at which the terrestrial vertebrates are adequately known). We can quite readily fit the distributions of the dinosaurs and other tetrapods of the Jurassic onto this model of Jurassic continents in the process of breaking apart. Laurasia and Gondwanaland were still recognizable, even though the elements that were to become our modern continents were taking form; rifting was taking place, but drifting had not as yet progressed to break connections between the several continental elements. Thus there were paths for faunal interchanges between the continents.

One final word. Although there are paths on this world model for land-living animals to move almost at will, there similarly are paths (except for Antarctica) on a theoretical Jurassic world with the continents in their present positions, with present or recent-past intercontinental connections. The difference between the two concepts, so far as the distributions of land-living vertebrates are concerned, is that on the world composed of a disintegrating Pangaea, the movement between Eurasia and North America was by way of a North Atlantic rather than a trans-Bering route, while the movement into Australia may have been through India rather than through the East Indies. Perhaps these differences appear to be academic. But perhaps they have real significance. The patterns of vertebrate distributions superimposed upon a disintegrating Pangaea fit nicely between the patterns of distribution in an ancient world of Gondwanaland and Laurasia (which already has been described) and patterns in the world that is more familiar to us and with which we soon will be concerned, namely, the world of Cretaceous time.

The Cretaceous world was not the world of today, but it was beginning to approach our modern world—geographically, climatologically, and biologically. It is to this world, when the dinosaurs reached the zenith of their evolutionary development in an environment that was in many respects modern, that we will now turn our attention.

VII. THE END OF AN ERA

CRETACEOUS SEAS AND LANDS

The name of the Cretaceous system of rocks comes from the Latin *creta,* which means "chalk"—a reference to the high chalk cliffs that bound a portion of the English Channel, where rocks of the Cretaceous system were first defined. The white cliffs of Dover are uplifted sediments that long ago accumulated on the bottom of a shallow sea; as such they represent a very important aspect of Cretaceous history. For during the Cretaceous period shallow seaways invaded the continents, as they had during the earlier phases of the Jurassic, but perhaps on a scale that previously had not been equaled. It was a time of marine transgressions, when tropical and subtropical seas were widely spread across the face of the earth.

Many of the Cretaceous seas were shallow and they extended in long, comparatively narrow arms across the still-drifting continental blocks. Although they divided the surfaces of the continents in varying patterns during the course of Cretaceous history, they did not affect the integrity of the drifting blocks. They must be thought of as evanescent surface features, comparable to the present-day Baltic Sea, or the English Channel, or Hudson Bay. These modern bodies of water have

FIGURE 68. The famous chalk cliffs of Dover, on the English Channel, from which the name of the Cretaceous period was derived.

loomed large as they have affected the march of human history, but to the geologist interested in the basic relationships of land masses they are in the nature of superimposed details.

Yet in spite of the wide extent of Cretaceous marine waters—both the deep oceans between the continental blocks and the shallow seas flooding portions of the several continents—the Cretaceous period was a time when there were broad expanses of land surface across which wandered the land-living vertebrates of that time. Not only were lands of the Cretaceous extensive, but also they were variously connected, so that, as in late Jurassic times, numerous tetrapods were able to move back and forth from one continental block to another, even though the continents were actively drifting apart. This should occasion no surprise. We now know that today the continents are drifting, probably as rapidly as they ever drifted, yet to us, as to countless millions of other animals inhabiting our modern continents, the lands seem as enduring as time itself. Geological processes are of such magnitude as compared with human experience that this is bound to be so. However, some of the violent earthquakes of the past few decades have served to remind us that the earth is far from being a static planet. The tectonic block containing much of California is moving northwest, scraping against the land mass to the east of it, often to the discomfort and sometimes the very survival of its residents. Only a few thousand years ago some of our ancestors walked from northern Europe into Britain on land that is now beneath the English Channel. Perhaps it will help to keep such homely examples in mind when we are discussing the changes in continental relationships through time.

EARLY CRETACEOUS LAND DWELLERS

The land-living vertebrates at the beginning of Cretaceous time were not very different from those that inhabited the late Jurassic continents. No great geological revolutions or events separated these two periods of earth history, and life continued, retaining much of its past quality. There were no profound extinctions among the land-living vertebrates during the transition from the Jurassic into the Cretaceous, as there had been from the Triassic into the Jurassic, or from the Permian into the Triassic periods, or, as we shall see, there would be when the Cretaceous period came to an end. Of course, many genera did not cross the time line between the Jurassic and the Cretaceous, as happened during all of the transitions between periods, and a few larger groups made their exit from the land of the living. For example, some families of crocodilians became extinct with the close of Jurassic time, but the crocodilians as such lived on with increased vigor, to become an even larger and more varied group of reptiles in the Cretaceous than they had been in Jurassic time. And the stegosaurian dinosaurs, so prominent among late Jurassic faunas, became extinct during Cretaceous time, although they did live into early Cretaceous years. One of the two suborders of pterosaurs disappeared at this time.

The Cretaceous was a time of certain innovations among the land-living vertebrates. Two important groups of dinosaurs, the armored dinosaurs and the horned dinosaurs, appeared during the Cretaceous, to become in some instances exceedingly abundant. The first snakes are found in Lower Cretaceous deposits. The birds, so primitive in late Jurassic time, rapidly evolved into essentially modern birds during the course of Cretaceous history. And this was the time when fur-bearing animals of modern aspect appeared in terrestrial vertebrate faunas, the time of early marsupial and placental mammals.

Thus, during early Cretaceous time the continents were variously inhabited by frogs and salamanders, turtles, some persistent eosuchians (in a sense, left over from Permo-Triassic times), rhynchocephalians, lizards, snakes, crocodilians, flying reptiles, all of the major dinosaurian suborders, birds, and marsupial and placental mammals. There is reason to think that the monotremes, represented today by the echidna and platypus of Australia, were also members of these early Cretaceous faunas, but for them there is no adequate fossil record. All in all this listing of animals, except for the dinosaurs and the flying reptiles, and those rare eosuchians, seems rather familiar. We are beginning to see outlines of the modern animal world.

THE WANDERINGS OF *IGUANODON*

It has been mentioned that continental connections were such during early Cretaceous times that many land-living vertebrates were able to move from one continent to another. This is nicely illustrated by the iguanodont dinosaurs (which are ornithopods) and especially by the genus *Iguanodon* itself. *Iguanodon* was one of the very first dinosaurs to be scientifically excavated and described, during the early years of the nineteenth century. Fragmentary remains were first discovered in southern England, and subsequently skeletons were found not only in England, but also and especially in Belgium, where in a coal mine a remarkable concentration of complete skeletons was excavated. This unusual assemblage of skeletons afforded paleontologists of the late nineteenth century perhaps their first detailed knowledge of a single dinosaurian genus. The fossils showed *Iguanodon* to be a large, heavily built ornithopod, yet a dinosaur that, in spite of its length of 40 feet or so and its probable weight in life of many tons, retained an ancestral bipedal pose. Needless to say, the hind legs were very strong. The forelimbs also were heavily built, even though they were relatively short, and each thumb terminated in a large, straight spike. The sides of the upper and lower jaws were well supplied with rather large, fluted teeth, together making a sizable grinding mill for crushing plant fibers, while in front the jaws were toothless and rather beaklike. In *Iguanodon,* however, the front of the skull and lower jaws had not as yet become the flat, ducklike bill that was to so characterize many of the later Cretaceous ornithopod dinosaurs. Thus *Iguanodon* became established as a characteristic European dinosaur. In more recent years fossils of this genus and of very closely related forms have been found widely distributed throughout the earth. Unmistakable iguanodont footprints were discovered, not many years ago, in Lower Cretaceous sediments on the island of Spitzbergen, now within about 12 degrees of the North Pole. *Iguanodon* is known from France, Portugal, and Spain, and, proceeding to the south, from the central Sahara region of Africa. Far to the east it has been found in the deserts of Mongolia. *Iguanodon,* or something closely akin to it, has recently been found in Queensland, Australia. And related forms are known from North America. This large dinosaur and its close cousins certainly must have moved from one region to another over dry-land connections.

The evidence of a single genus such as this is in a way dramatic and certainly convincing. But there is a broader base than that of the iguanodonts upon which to view the distribution of land-living tetrapods within the continental regions of early Cretaceous times. The comparisons of several tetrapod faunas of this age are revealing.

Iguanodon, present in England, Belgium, France, Spain, and Portugal, is but one element, albeit a spectacular one, of the Wealden fauna, which inhabited northern Europe during the early part of Cretaceous time. This fauna is named from the Wealden beds, typically exposed in the Weald of southern England—the name "weald" being of

FIGURE 69. The distribution of iguanodont dinosaurs at the advent of Cretaceous history. From what is now the island of Spitzbergen to Australia, these dinosaurs were distributed through wide ranges of latitude—an indication of rather uniform world climates during late Mesozoic history.

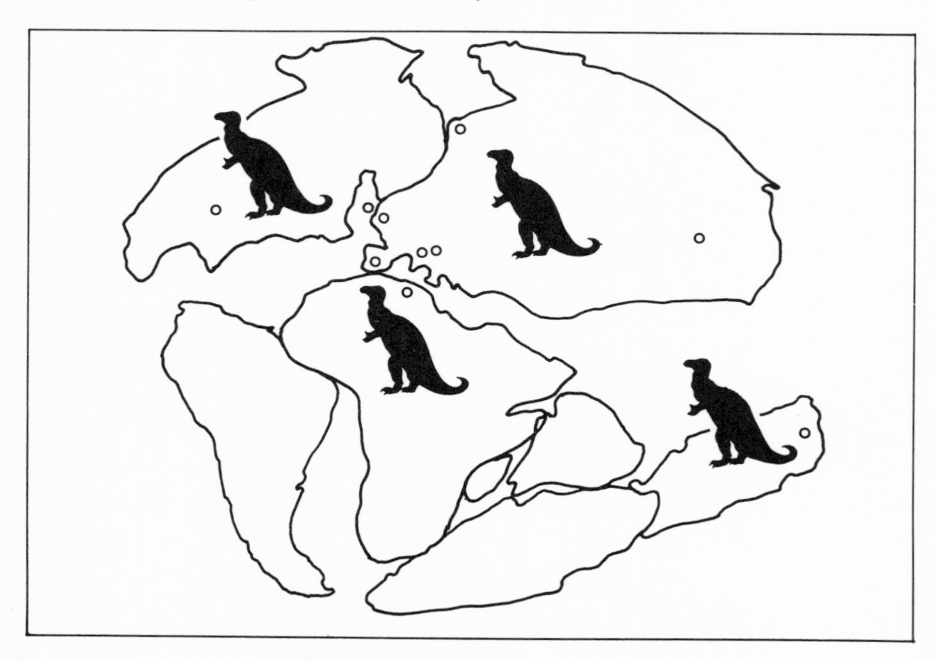

Anglo-Saxon and Middle English derivation, and meaning an "open country," or "wold." It was in the open country of fields and woods bordering the English Channel that *Iguanodon* was first discovered, and it was just across the Channel, in Belgium, that the remarkable series of skeletons of this interesting dinosaur was unearthed.

LOWER CRETACEOUS DINOSAUR FAUNAS AND CONTINENTS

The Wealden fauna is well documented and diverse. It contains various fishes (which do not concern us here), reptiles, and some mammals. The reptiles consist of turtles, crocodilians, gigantic sauropod dinosaurs, coelurosaurs and carnosaurs (representing the small and large carnivorous dinosaurs), a small, primitive ornithischian known as *Hypsilophodon* (a sort of ancestral edition of *Iguanodon*), *Iguanodon* of course, several armored dinosaurs, and a probable stegosaurian. The mammals in the fauna are present as three very primitive genera, all very small. There also very probably were the various small tetrapods in the fauna that all too often are not preserved in the sediments—animals such as frogs and lizards.

In Montana and adjacent states is exposed the Lower Cretaceous Cloverly Formation, containing a tetrapod fauna that nicely parallels the Wealden fauna. There are no generic identities, but most of the same groups comprising the Wealden fauna are present in the Cloverly association, as closely related genera. The Cloverly fauna is deficient in fishes, as contrasted with the Wealden, but it has turtles, crocodilians, and several groups of dinosaurs, similar to those of the Wealden. No mammals have been found, but this very probably is an accident of preservation or of collecting, or both.

A Lower Cretaceous formation, the Arundel, on the eastern seaboard of North America also contains a rather fragmentary fauna comparable to the Wealden, it being typified by giant sauropod, carnivorous theropod, and armored dinosaurs, crocodilians, and turtles. Again, as in the case of the Cloverly fauna, there are no generic identities with the Wealden.

Even though these North American Lower Cretaceous faunas are different in detail from the Wealden assemblage, they are sufficiently close to show a community of origin, and that not very far removed in time. In other words, there must have been land connections between the two continental regions in order that such similar faunas would have developed.

In the other direction, in Mongolia and China, are Lower Cretaceous tetrapod faunas that parallel the European Wealden fauna. Here are found small carnivorous dinosaurs, large sauropods, iguanodonts and armored dinosaurs, and crocodilians. In at least one instance there is a generic identity between the Asiatic faunas and the type Weald fauna, for *Iguanodon* has been recorded in Mongolia.

Looking to the south, there are Lower Cretaceous faunas in Africa that show even closer resemblances to the European Wealden fauna than do the contemporaneous assemblages of Asia and North America. In Morocco there are the genera *Megalosaurus,* a gigantic carnivore similar to the large carnivorous dinosaurs already described, and *Cetiosaurus,* a large sauropod, both of which are typical of the European Wealden. In the central Sahara are small and large carnivorous dinosaurs, sauropods, stegosaurians, and the genus *Iguanodon.* At the southern tip of the African continent are the Uitenhage beds, which contain fragments of Lower Cretaceous dinosaurs.

Discoveries of dinosaurian remains in the Nequen beds of Argentina and in eastern Australia (the presence of *Iguanodon* in this region already has been mentioned) complete the picture, showing a worldwide distribution of land-living tetrapods during the early part of Cretaceous history. Evidence from peninsular India is as yet lacking. In some cases, such as those just mentioned, the evidence is not as yet abundant, but it is there. And it does show that land-living vertebrates inhabited virtually all of the continental blocks, which then were in the course of fragmentation and drift from the parent Laurasian and Gondwanaland continents.

How does the distribution of land-living vertebrates of Lower Cretaceous age accord with the continental relationships as they are postulated for this phase of geologic history? First it is necessary to look at the continents of early Cretaceous time briefly, but nevertheless in somewhat greater detail than they were outlined in preceding paragraphs, in order to obtain the geographic background for the Lower Cretaceous faunas.

As has been said, even though the early Cretaceous was a time when continental areas were of such magnitude and connected in such fashions as to give scope not only for the development of varied tetrapod faunas, but also for the close intercontinental relationships of these faunas, it was still a time when seas had spread widely across the face of the globe. Moreover, the spread of these seas was in part the result of the flooding of continental surfaces by the incursions of shallow waterways. Naturally such developments on the surface of the globe affected the course of biological evolution.

It would appear that Laurasia continued to be essentially a unit, the North American–Eurasian ligation persisting across the Greenland area. This might account for the evident relationship of the Wealden and the Cloverly-Arundel faunas. There is good reason to think, however, that during early Cretaceous time an arm of the sea was encroaching from the north over the surface of northwestern Canada while at the same time an embayment was developing from the south, from the Gulf of Mexico in a northwestwardly direction. Thus, although North America persisted as the western block of Laurasia, its surface was partially divided into eastern and western portions. It is possible that the evolving tetrapods of western North America were partially blockaded from those of the eastern edge of the continent, and from those of the Wealden region. (Later in the Cretaceous, as we shall see, the marine invasions extended across the continent, from north to south.) If this were so, it might have brought about the closely related but independently parallel development of the faunas—so that generic distinctions were the rule.

In a similar manner narrow seas were encroaching upon Asia from the north and the south, to the east of the Ural Mountain region. In early Cretaceous time the eastern part of Asia probably was not completely cut off from western Eurasia, as it was to be in the later phases of Cretaceous history, so that there was a broad area in existence that allowed east and west movements of land-living tetrapods. This presumed corridor may explain the presence of *Iguanodon* in eastern Asia. As for other reptiles contemporaneous with this dinosaur in eastern Asia, perhaps time and distance had brought about the generic distinctions that we see between them and the Wealden reptiles. Perhaps there was even a certain amount of filtering of faunal elements in this part of the world; *Iguanodon* made the trip but other Wealden genera did not. However that may be, certainly there was some land connection between eastern and western Eurasia.

The marine waters that invaded Eurasia from the south during early Cretaceous time were an extension of the ancient Tethys Sea, which had for so long formed a great east and west ocean between the Eurasiatic part of Laurasia and the African part of Gondwanaland. These waters flooded much of eastern Europe, but on the west there were connected lands, from North America, through England, down through France, Spain, and Portugal, and, it would seem, into Africa.

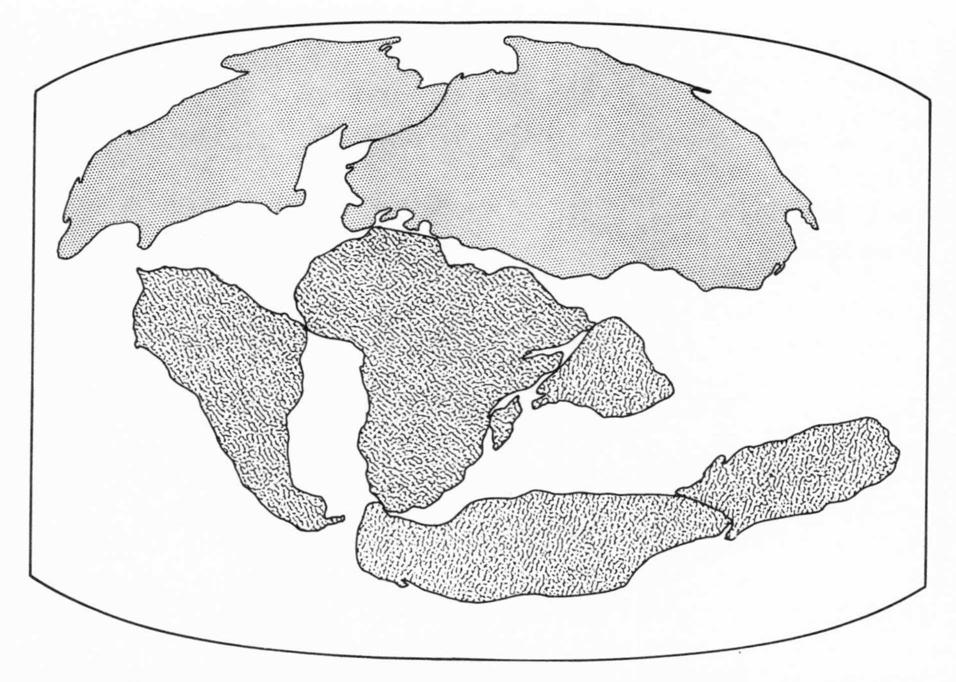

FIGURE 70. A reconstruction of Pangaea during Cretaceous times. Large dinosaurs in South America, India, and Australia suggest the persistent connections of these land masses (the last through an Antarctic ligation) to Africa.

This is the region where the Wealden fauna as such flourished, with *Iguanodon, Megalosaurus,* and *Cetiosaurus,* to mention three very typical Wealden dinosaurs, extending southward into the heart of Africa.

The presence of dinosaurs in South America and in Australia, an iguanodont having reached Queensland, as we have seen, indicates land connections to these two continental regions. Some authorities would have the South Atlantic Ocean still very narrow at this stage of earth history, so that a northern connection between South America and Africa might still have been in existence. If so, a route for the interchange of land-living animals would have been available. Also, some authorities would place the separation of the Indian-Australian-Antarctic block of Gondwanaland from the African block as an event of early Cretaceous date. Consequently there might have been a route to Australia from Africa that would have allowed the invasion of iguanodonts into this part of the world. If, on the other hand, the separation of India and the Antarctic-Australian block had occurred at a somewhat earlier date, as has been suggested in the discussion of Jurassic continents and tetrapods, then the presence of dinosaurs in Australia is not so easy to explain—and India becomes an enigma. One cannot be dogmatic about these problems of continental relationships and faunal distributions in the far distant past. It must be recognized that our view of the earth and of life in those distant epochs of earth history is still incomplete and fragmentary, and we must wait for time and future discoveries to clarify the picture.

On the whole, however, the distributions of Lower Cretaceous land-living tetrapods accord reasonably well with the theory of continental drift.

A NEW WORLD OF LATE CRETACEOUS FLOWERING PLANTS

During the course of early Cretaceous history—and certainly by the beginning of late Cretaceous time—there had occurred an evolutionary development that was to influence profoundly the faunas of land-living vertebrates. This development occurred in the plant kingdom; it was the rise and the rapid evolutionary radiation of the angiosperms, the flowering plants.

Consequently late Cretaceous landscapes assumed a modern and familiar appearance, with oaks, sassafras, and other such hardwood trees filling woodlands that covered rolling hills, with willows along streams and with flowering magnolias adding color to the scene at certain times of the year. There were many other hardwood trees, too numerous to mention, and many flowering plants. The vistas were varied and colorful in all directions—a distinct contrast to the green monotony of the earlier Mesozoic forests, in which conifers were the dominant trees. Of course, the burgeoning of flowering plants during the later years of Cretaceous history was accompanied by the correlative increase of insects—the carriers of pollen from plant to plant. And it was accompanied, also, at the opposite scale of size among land-living animals, by a remarkable increase in the plant-eating dinosaurs, notably the ornithischians. A new food supply had come into being, with the result that new herbivorous dinosaurs arose and evolved, to feed upon this source of energy.

THE ZENITH OF DINOSAURIAN EVOLUTION

Among the dinosaurs that so developed, in many cases from early Cretaceous ancestors, were the numerous and highly varied hadrosaurs, or duck-billed dinosaurs, representing the culmination of ornithopod evolution. And with them there evolved the curious pachycephalosaurs, or dome-headed dinosaurs, in which the top of the skull was thickened into an immense bony boss above the brain. Also the ankylosaurs, or armored dinosaurs, which had already made their appearance in early Cretaceous time, reached the zenith of their evolutionary development. Finally there was a new group of Cretaceous origin and evolution, the ceratopsians, or horned dinosaurs. These reptiles may have appeared during the early Cretaceous, but the time of their major evolutionary radiation was the late Cretaceous, when they became horned giants, fulfilling the role in the late Cretaceous landscapes that the rhinoceroses were to assume in the Cenozoic and modern world. These last of the large dinosaurian groups to arise and evolve would seem to have been limited to western North America and northeastern Asia, a point that will be explored in more detail later.

Certainly the dinosaurian herbivores were numerous, according to the abundance of their fossil remains, and in this respect were, of course, fulfilling the role of primary feeders upon an abundant food supply—as do, for instance, the various antelopes in modern Africa. Living with the duck-billed, armored, and horned dinosaurs were giant theropods, the predators, the secondary feeders, which depended upon the herbivores for their food supply. It is no wonder that the plant-eating ornithischians of the Upper Cretaceous were so efficiently protected against their gigantic predatory neighbors—by escape to the water in the case of the duckbills, by heavy armor in the case of the ankylosaurs, and by actively fighting back in the case of the horned ceratopsians. Such defenses were necessary to survive.

As for other dinosaurs of late Cretaceous age, there were some surviving gigantic sauropods, although they did not dominate the faunas as they had in late Jurassic time. And there were small, lightly built theropods that evidently preyed upon small animals or perhaps, in some cases, fed upon fruits and the like.

All in all, the end of the Cretaceous was in many respects the culmination of dinosaurian evolution. Never before had dinosaurs been so varied; never had they shown such a wide range of adaptations for different modes of life. Very possibly never before had they been so numerous. It would seem that these remarkable reptiles had attained a level of evolutionary success that should have made them masters of the continents for uncounted geologic ages into the future. Yet they were destined soon to disappear from the face of the earth. In the meantime, however, they dominated the lands in a fashion that probably had

FIGURE 71. Late Cretaceous dinosaurs of North America. In the right foreground are the duck-billed dinosaurs *Parasaurolophus*, and on the extreme left the gigantic predator *Gorgosaurus*, the largest known land-living carnivore. Between them is armored dinosaur *Palaeoscincus*, and just beyond another duck-billed dinosaur, *Corythosaurus*, the nearest one of which is looking toward the horned dinosaur *Styracosaurus*. In the middle distance, standing on the shore, are the lightly built theropods *Ornithomimus*. Life was varied by the end of Cretaceous times, as was the physical world; it was the time when modern mountain systems were coming into being.

never before been reached by the vertebrates. They wandered far and wide across the Cretaceous continents. Moreover, it is interesting to note that they wandered through landscapes of very modern aspects.

BIRDS AND MAMMALS OF THE CRETACEOUS

The backboned animals that shared the lands with these last of the dinosaurs were for the most part animals having an appearance of modernity, in keeping with the colorful flowering plants and broad-leaved trees around them. Modernity for the most part, but not entirely, because the pterosaurs, or flying reptiles, still inhabited the earth. In some regions, as for instance around the shallow sea that covered middle North America, the air was the province of pterosaurian giants, such as *Pteranodon,* with leathery wing spreads of more than 25 feet. Moreover, a few persisting eosuchians still inhabited the land—long-snouted lizardlike reptiles that may have lived lives somewhat between those of large lizards and small crocodilians.

Otherwise, however, the land-living vertebrates were of familiar form. There were frogs and salamanders in moist places. Turtles lived in streams and lakes and on the land as they do now, and there were crocodilians widely spread across the continents. Lizards and snakes were present, as were rhynchocephalians. Although the air was the home of the flying reptiles, birds were now abundant, and perhaps owing to the competition from these very efficient flying vertebrates the pterosaurs eventually became extinct. Some birds had already specialized away from the pattern of flight, as for example *Hesperornis,* a very loonlike but wingless bird that lived along the shores of the mid-American Mediterranean-type sea and swam and dived in pursuit of fishes, as does the modern loon. Finally, the earth was the home of numerous mammals—marsupials, some of which were very similar to the modern North American opossum (which is in essence a persisting Cretaceous mammal), and small placental mammals as well. These last, tiny insectivores related to the modern shrews and hedgehogs and quite inconspicuous in faunas that were dominated by the many dinosaurs, were of very great significance, because from them arose the host of placental mammals that during Cenozoic times inherited and dominated the earth. We are among this host.

There were in addition some rather rodentlike mammals known as "multituberculates," a group that flourished during Jurassic and Cretaceous times and persisted into the beginning of the Cenozoic era, only

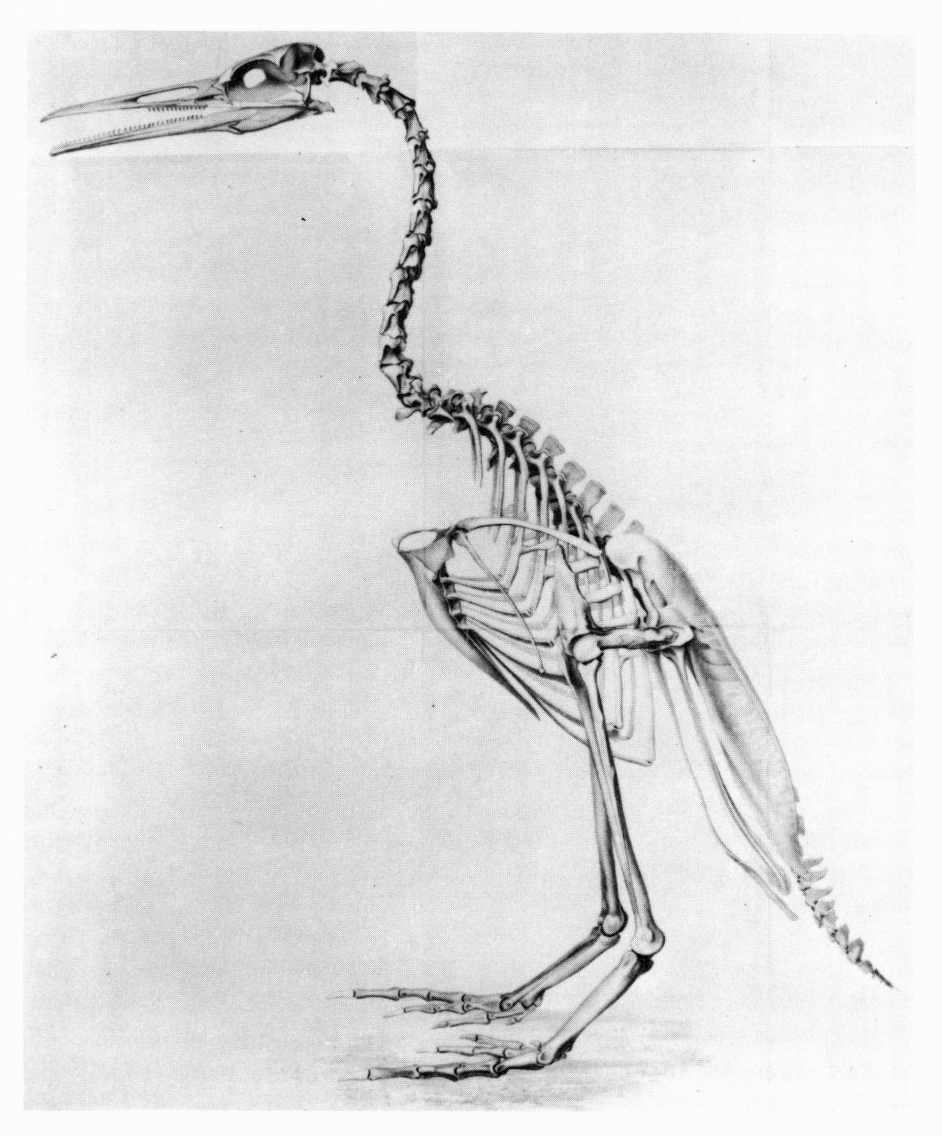

FIGURE 72. *Hesperornis,* a wingless, loonlike bird that inhabited the shallow, mid-continent Cretaceous ocean of North America.

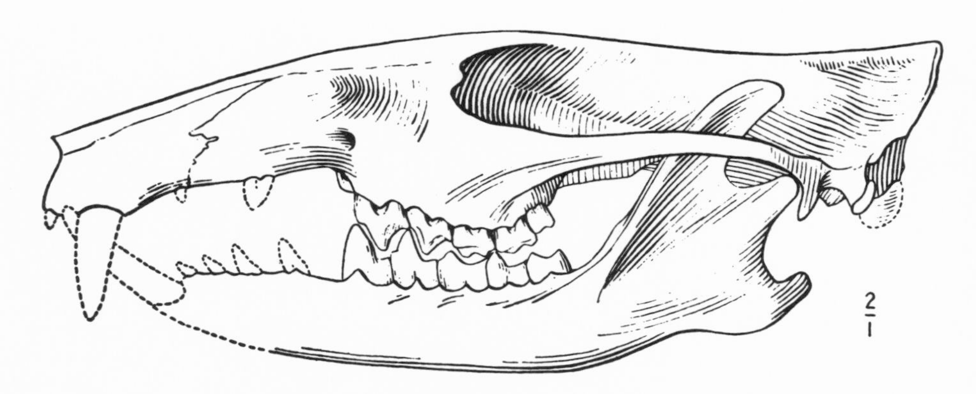

FIGURE 73. An insectivore (*Zalambdalestes*), one of the earliest known placental mammals, from the Cretaceous beds of Mongolia. This little skull, about two inches in length, represents an animal that seems inconspicuous when compared with the giant dinosaurs of that time. But from an insectivore stem evolved the varied host of modernized mammals that came to rule the world after the extinction of the dinosaurs.

to become extinct, probably in the face of competition from the placental rodents. These interesting mammals were characterized by large, pointed or chisellike incisor teeth in the front of the skull and lower jaw, and by elongated cheek teeth in which there were longitudinally arranged rows of cusps. Frequently there was a large bladelike cutting tooth, with a fluted outer surface, on each side of the lower jaw. Such an arrangement of teeth was obviously well suited for reducing husked fruits and perhaps nuts to edible proportions, yet even so the true rodents, with somewhat similar tooth mechanisms, but probably with a more efficient reproductive system, eventually prevailed. Such is an outline of tetrapod life on the continents as the Age of Dinosaurs drew to a close.

LATE CRETACEOUS CONTINENTS

At this time the drifting continents were coming ever closer to their modern positions. The world was becoming modern in aspect, yet it still retained some of its ancient heritage. The long-established northern connection between northeastern North America and northwestern Europe would seem to have persisted as a ready route for the exchange of land-living vertebrates on a Laurasian continent. The North Atlantic Ocean, becoming wider, was nevertheless closed at its northern end. On the other side of the globe at these northern latitudes it would appear that there was a land bridge between northwestern North America and northeastern Asia. This was the trans-Bering bridge, which was to continue intermittently from that day to this. Today the Bering Strait separates North America from Asia, but this is indeed a narrow and shallow seaway, barely 50 miles wide and 150 feet deep. Eustatic (or vertical) movements of no great magnitude could again restore the Bering land bridge, linking the northern-hemisphere land masses as they have so often been connected in the past.

Although there were connections joining eastern North America with Europe and western North America with Asia, there were then narrow north and south seaways across North America and Asia that bisected the surfaces of these continents, at least for portions of late Cretaceous time. It will be recalled that in discussing the paleogeography of the early Cretaceous, mention was made of the oceanic embayments that were encroaching upon the surfaces of Asia and North America, in Asia to the east of the present Ural Mountains, in North America in the midcontinent region to the north of the Gulf of Mexico. The evidence indicates that the process of encroachment, so apparent in deposits of early Cretaceous age, was brought to its culmination in

FIGURE 74. A multituberculate, an early experiment in mammalian evolution that failed. The multituberculates (so named because of the several cusps or tubercles on the crown of each cheek tooth) arose during the age of dinosaurs and lived quite successfully through the years of the giant dinosaurs and into the beginning of the age of mammals. It would seem that perhaps they occupied for a time the role of rodents. But when the true rodents appeared, the multituberculates quickly died out, seemingly unable to withstand rodent competition.

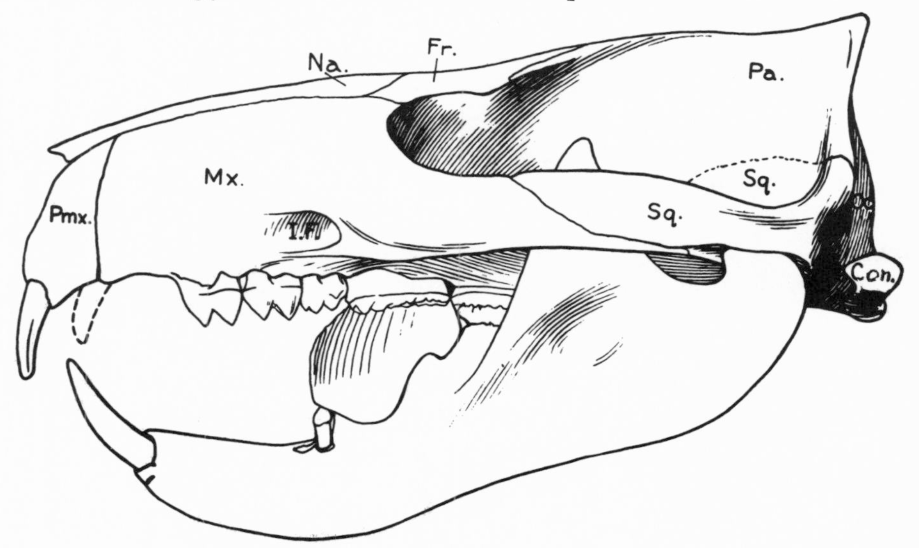

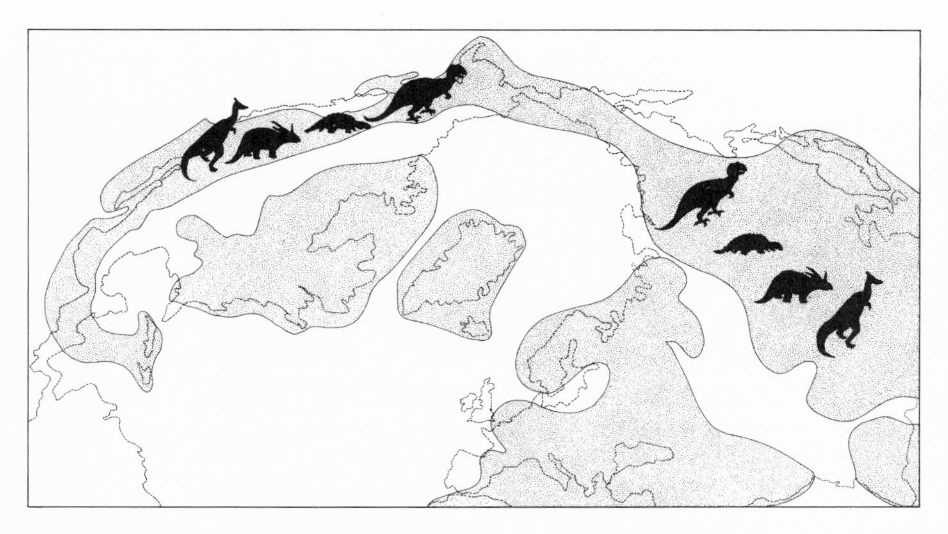

FIGURE 75. An interpretation of Laurasia during late Cretaceous history. The presence of essentially identical dinosaurs in Mongolia and in western North America is strong presumptive evidence favoring a trans-Bering bridge that would permit the movements of these giant reptiles back and forth between the two areas. Perhaps the extension of shallow, north-to-south Mediterranean seas across the middle of North America and of Eurasia served during some of Cretaceous time as barriers limiting the distributions of these dinosaurs.

the late Cretaceous. Consequently eastern North America and western Eurasia formed one land unit, joined by the northern Greenland land bridge and only partially separated by a narrow North Atlantic Ocean, while western North America and eastern Eurasia was another land unit, joined by the northern trans-Bering land bridge. These two lands were separated from each other on the surface by the North American midcontinent sea and by the Asian continental sea, both of late Cretaceous age. One must not forget, however, the fact that, although land-living animals were in many instances restricted by these longitudinal seaways, the continental blocks as such remained intact—the North American block ever drifting to the west, straining to break the final connection (which was to become Greenland) that still bound it to Eurasia. Laurasia was still in existence, but in modified form, and it was not to remain as one unit for much longer, geologically speaking.

It seems probable that the westward drift of South America had by now proceeded to such a stage that this continental block was free from Africa. If so, the South Atlantic probably was still relatively narrow. And if so, South America still must have retained some connections with other continental regions, as is indicated by the skeletons of dinosaurs and other reptiles from the Upper Cretaceous of Brazil and Argentina. It seemingly had not as yet become completely an island continent. Perhaps there was even a connection in the form of a Panamanian isthmus, newly established, to link the westward-drifting South American block with North America.

The ancient link between Gondwanaland and Laurasia still may have been in effect during late Cretaceous time, either in the western Mediterranean region or in the eastern portion of the Mediterranean. Dinosaurs are found in Egypt and in Morocco, as testimony of the close faunal relationship of this part of Africa with lands to the north. And it seems likely that land-living vertebrates related to those of northern Africa inhabited a large part of the continent during late Cretaceous time.

It will be remembered that large dinosaurs of early Jurassic age are known from peninsular India, being found in the Kota Formation of that region. The record of land-living vertebrates in peninsular India would seem to be a blank until we reach the Upper Cretaceous, when again dinosaurs are found—in the Lameta beds. Dinosaurs have also been found in the Upper Cretaceous beds of eastern Australia.

Such is our view of the earth toward the end of the Cretaceous period, when the continental blocks were approaching their present positions. Some of the Cretaceous relationships between continents adumbrated those characteristic of later times, while some retained ancient connections. To summarize, it is suggested that eastern North America, connected to Europe by a northern passage, was separated from western North America (at least during a part of the Cretaceous) by a longitudinal seaway, while western North America was connected to eastern Asia by a trans-Bering bridge, and this latter land area was separated from western Eurasia by another longitudinal seaway. If South America had by this time become completely separated from Africa, it probably had connections with the rest of the world, perhaps by a Panamanian isthmus. Africa still retained connections with Laurasia to the north. Australia may have had connections through Antarctica to Africa.

DISTRIBUTIONS OF UPPER CRETACEOUS DINOSAURIAN FAUNAS

We may now turn our attention to Upper Cretaceous land-living vertebrate faunas, to see how well they accord with this reconstruction of the late Cretaceous earth.

Perhaps the most spectacular and abundant of Upper Cretaceous dinosaurian faunas (using the term "dinosaurian" here to indicate the dominant elements in the faunas, but realizing, once again, that such faunas contained various other vertebrates as well) are those of western North America and eastern Asia. In western North America these faunas are found in closely related Upper Cretaceous formations, from Alberta and Montana to New Mexico and Mexico. Such are the faunas, from north to south, of the Lance, Hell Creek, Edmonton, Kirtland, Fruitland, Oldman, Milk River, Judith River, Two Medicine, Mesa Verde, Aguja, and Difunta formations. In these beds are found hosts of dinosaurs, among which the duck-billed, armored, and horned dinosaurs are the prominent herbivores, and among which the carnivores are represented by gigantic theropods. There are also some large sauropod dinosaurs in some of these formations. Various other tetrapods are present, represented by turtles, lizards, and snakes, crocodilians, eosuchians, rhynchocephalians, and primitive mammals. Flying reptiles and birds are found in related sediments, especially the Niobrara chalks of Kansas.

FIGURE 76. Boxing a skeleton of the giant carnivorous dinosaur *Tyrannosaurus* in the Cretaceous Hell Creek beds of Montana.

Paralleling these dinosaur-bearing sediments are the dinosaur beds of the Mongolian Gobi—the Bain Shire, Nemeget, and Djadochta (or Bain Dzak) formations, and more or less correlative sediments in Kansu, Shantung, and Sinkiang. The Lameta beds of peninsular India, already mentioned, may be equated with the Upper Cretaceous beds of Mongolia. It is most interesting that the vertebrates occurring in these northern Asiatic sediments are very closely related indeed to those found in the Upper Cretaceous formations of western North America. To set down the roster for one area is in effect to set it down for the other; so the tetrapods listed above as characteristic of the western North American Cretaceous are equally characteristic of the eastern Asiatic Cretaceous. In many instances the similarities extend down to the generic level. A *tyrannosaur,* the gigantic predator from the Hell Creek beds of Montana, a dinosaur standing 20 feet or more in height, measuring 40 feet or more from nose to tail, having an enormous skull armed with large, bladelike teeth, and characterized by ridiculously small forelimbs (the function of predation evidently being concentrated in the huge jaws), likewise is found in the Nemeget sediments of Mongolia. The duck-billed dinosaur *Saurolophus,* a large ornithopod, in which the skull and lower jaws were flattened in front, like the bill of a giant duck, while at the back of the skull a large spike protruded up and to the rear, is found in both regions. So are the horned dinosaurs *Styracosaurus* and *Pentaceratops,* the former with a huge skull, fully one-third of the total length of the animal, the back portion of the skull being in the form of a broad frill, the edges of which were decorated with long spikes, while on the nose there was a long, straight horn; the latter with an equally large skull, the frill of which was not so elaborately decorated, but in which animal there were three horns—a short one on the nose and two long horns, one above each eye. Such identities, plus the very close relationships evident among other forms that are not generically the same, must indicate an intimate connection between North America and Asia. And this connection must have been a trans-Bering ligation.

It may be significant that horned dinosaurs, which occur in western North America and in eastern Asia in abundance and variety, have not been found elsewhere. Of course, future discoveries may change the picture, but at the present time the confinement of these dinosaurs to the areas indicated appears to represent the true limits of their ranges, especially since sediments in other regions that logically should contain such fossils have not as yet yielded them. One is tempted to think that the geographic restriction of the horned dinosaurs is owing to the

FIGURE 77. Cretaceous beds in Mongolia, with the camp of the Polish-Mongolian paleontological expedition in the foreground. In the sediments of the Nemegt basin was found the skeleton of a tyrannosaur (*Tarbosaurus*) very closely related to the tyrannosaur from the Hell Creek beds of Montana.

barriers of the North American midcontinent Cretaceous seaway on the east and the mid-Asiatic seaway on the west.

That eastern North America was not completely isolated from western North America during late Cretaceous time is shown by the presence along the Atlantic seaboard of duck-billed dinosaurs as well as some of the smaller carnivorous theropods. It is likely that there were land connections at some intervals during the late Cretaceous, but not at such times or in such forms as to allow the upland-living horned ceratopsians to make the journey to the east.

The nature of carnivorous dinosaurs, duck-billed dinosaurs, other dinosaurs, and tetrapods other than dinosaurs demonstrates the continuation of the eastern North American–western European ligation in late Cretaceous time. European dinosaurs of this age are represented in the Upper Cretaceous of Transylvania.

In Africa Upper Cretaceous dinosaurian faunas are especially well represented along the northern border of the continent, in Morocco and in Egypt. The discoveries in Morocco are in the form of well-preserved trackways. In Egypt is the Baharia fauna, from the sediments of that name, characterized by gigantic carnivorous dinosaurs (one of which, *Spinosaurus*, is distinguished by the enormously long spines of the vertebrae) and by sauropods. The record of Upper Cretaceous dinosaurs extends well down into the central part of the continent, as a result of work carried out during recent years.

France has been especially productive in recent years, and a considerable group of Upper Cretaceous dinosaurs is known from this region. Included are large carnivorous dinosaurs, sauropods, advanced iguanodonts, duck-billed dinosaurs, and armored dinosaurs. It is a characteristic Upper Cretaceous assemblage, and though generically distinct from similar dinosaurs found in North America, it nevertheless indicates dry-land communication between this part of Europe and the North American region during the latter portion of Cretaceous history. Of particular interest has been the discovery of large numbers of fossilized eggs, presumably dinosaurian, in the Upper Cretaceous deposits near Aix-en-Provence, in southern France.

The connection between South America and other continental regions in late Cretaceous time has been mentioned. There are two areas on the South American continent where land-living vertebrates of this age have been found—Brazil and southwestern Argentina. In Brazil is the Bauru Formation, containing not only large dinosaurs, but also a remarkable array of fantastic crocodilians and gigantic turtles. Here is a record of life which, although in obvious contact with the rest of the world, as shown by the dinosaurs, was still sufficiently isolated for evolution among some animals, such as the crocodilians, to proceed along lines quite different from anything seen elsewhere.

The Upper Cretaceous dinosaurs of Argentina are sauropods that would seem to show relationships to Upper Cretaceous sauropods in India. For example, the genera *Titanosaurus* and *Laplatasaurus*, generally similar to other sauropod dinosaurs (all of the sauropods were cut to a rather uniform pattern) but sufficiently distinct in certain details to be designated as separate genera, both typical of the Argentinian deposits, have been described from the Lameta beds of India. *Laplatasaurus* has also been assigned to the Upper Cretaceous of Madagascar. If these genera are truly in the localities indicated, we have a record either of long intercontinental movements, as would be necessary if the drift of continents had progressed as far as has been indicated in the preceding discussion, or of a relatively straight-line distribution from South America through Africa and into peninsular India on an as yet partially intact Gondwanaland. The dilemma in this case is one of opposing judgments. Are the various evidences of geology, geophysics, and paleontology (other than of the sauropod dinosaurs) to be accepted, thus indicating an advanced stage in continental drift, or are reputed distributions of these particular sauropod dinosaurs to be taken as an indication that parts of Gondwanaland were still connected—South America to Africa and Africa to Madagascar? It is here suggested that the weight of the evidence favors the first interpretation —namely, that the fragmentation of Gondwanaland and the subsequent drift of its several parts were well advanced. The whole question as to the distribution of the late Cretaceous sauropod dinosaurs needs to be reexamined, with perhaps a new and a skeptical look at the identification of the specimens found in Argentina, Madagascar, and India. Perhaps endemic genera are represented in these now widely separated localities. It should be said in this connection that the identification of sauropod dinosaurs, especially when skulls are lacking, can be remarkably difficult. All in all, this one aspect of late Cretaceous continental relationships and tetrapod distributions must for the time being remain an open question.

Dinosaurs in the Upper Cretaceous of eastern Australia have been mentioned. As in the case of the Lower Cretaceous iguanodonts from Queensland, such occurrences indicate the continued connection of Australia with other continental areas.

The distributions of land-living tetrapods in Upper Cretaceous deposits on the whole accord reasonably well with the intercontinental relationships predicated upon the continued drift of continental blocks from a fragmented Pangaea. Certainly the relationships of most of the Upper Cretaceous faunas in the northern hemisphere can be fitted into a Laurasian block, the eastern and western segments of which were bisected by north and south seaways. And there is no problem in relating Africa to this northern land mass. The problems concern peninsular India, Australia, and South America, all of which obviously had connections with other parts of the world. Where these connections were located are matters for speculation.

However that may be, it would seem that, generally speaking, the world by the close of Cretaceous time was approaching its present geographical lineaments. As we have seen, the plant world of this time was essentially modern in aspect, even though the continued presence of giant dinosaurs gave landscapes an ancient look.

THE END OF THE CRETACEOUS

But with the end of the Cretaceous, the end of the dinosaurs was at hand. During the transition from Cretaceous to Cenozoic time the

FIGURE 78. Upper Cretaceous dinosaurs from North America and Asia. *Tyrannosaurus* (left) as displayed at the American Museum of Natural History in New York and *Tarbosaurus* as displayed at the Municipal Museum in Ulan Bator, Mongolia. It seems obvious that dinosaurs of this type ranged back and forth across a trans-Bering bridge during the final years of Cretaceous history.

dinosaurs disappeared, and geologically speaking they vanished rather suddenly. Why the dinosaurs should have become extinct at the end of Cretaceous time is one of the great puzzles of geologic history. Theories almost without end have been proposed to explain the passing of these great reptiles—theories that range from the persuasive to the ridiculous. Yet every theory has its flaws, and for every explanation setting forth the reasons for the disappearance of the varied dinosaurs exceptions and objections can be brought forward. In short, why did all of the dinosaurs die out at the end of Cretaceous history? Why did not some of them survive, as did their close cousins the crocodilians? We are puzzled by this phenomenon—but the hard fact remains, the dinosaurs completely vanished from the face of the earth as the transition took place between Cretaceous and Cenozoic time.

It is said that nature abhors a vacuum; certainly the demise of the dinosaurs opened the surfaces of the continents to domination by new animals, and these new animals were the mammals. From the tiny insectivores of the Cretaceous the placental mammals exploded, in the evolutionary sense and with the advent of the Cenozoic, to populate the world with strange beasts, some small, some large. These new ruling mammals moved into the lands left vacant by the disappearing dinosaurs in patterns that were determined by the ecological niches available and by the interrelationships between the still-shifting continental blocks.

The patterns of continental relationships at the beginning of Cenozoic time were obviously inherited from the late Cretaceous patterns of arrangements of continents and continental connections. And, of course, the early mammals moved from one region to another according to the distributional lanes that were open to them. Their early evolution and their early domination of the lands were thus determined in large part by continental positions and connections. As the relationships of the continents to each other gradually changed during the course of Cenozoic history, the fortunes of the mammals on the continents also changed. How these changes took place is a story rather different from the long history of continents and animals with which until now we have been engaged. It is a story of the world finally coming into its modern condition—geographically, climatologically, and biologically. It is now our purpose to follow the fortunes of the new land-living vertebrates, primarily the mammals, in this changing world.

VIII. PRELUDE TO THE MODERN WORLD

THE MESOZOIC–CENOZOIC TRANSITION

Although the history of Cenozoic land-living mammals is rather different from that of the Mesozoic terrestrial amphibians and reptiles that preceded them, owing, of course, in part to the changing relationships between the continental masses, the difference is not a matter of sudden and sharp distinctions between the lands and the animals of the two geologic eras. Changes in the lands of the earth and among the plants and animals living on these lands have been on the whole gradual processes throughout geologic history. Perhaps this does not always seem to be the case when we look at the geologic record, because the record is at best imperfect, and events have the appearance of being telescoped through the passage of time. Yet we know that time has been a continuing process, that the years have passed by inexorably, carrying with them the changing fortunes of life on the earth.

So it was that the Mesozoic era, the great Age of Dinosaurs, passed into the Cenozoic era, the age of mammalian dominance. And so it was that these two disparate patterns of ruling vertebrates on the land reflected not only the evolutionary differences that distinguished Mesozoic and Cenozoic life, but also the differences in the relationships of the continents. The differences, however, did not become immediately established; in many respects the beginning of the Cenozoic world was very much like the end of the Mesozoic world.

This is certainly true with regard to continental relationships. The drifting continents continued to drift, and those connections that have been outlined as characteristic for the final days of Cretaceous history continued into the early days of Cenozoic history. Moreover, environments of the early Cenozoic would seem to have continued those environments that had prevailed during late Cretaceous time. The early Cenozoic world was on the whole benign, with tropical and subtropical temperatures of wide extent. The distinct zonation of climates so familiar to us probably was still largely undeveloped. There certainly must have been a gradient in climates and in temperatures from the equator to the poles, but the gradient seemingly was less extreme than it is today. The tropical and subtropical belts were wider than they now are; temperate climates extended into high latitudes, and it would appear that there were no extensive polar ice caps.

Furthermore, the plant life of early Cenozoic time was a continuation of that typical of the late Cretaceous. We have seen how a revolution in plant life had occurred during early Cretaceous time, when the flowering angiosperms spread across the continents to colonize lands that heretofore had been clothed with gymnosperms (such as various conifers) and with more primitive plants (such as ferns). We have seen that the last of the dinosaurs lived in forests and savannas of modern aspects. Indeed, it was the advent of the flowering plants that very probably determined the remarkable diversity among the late Cretaceous herbivorous dinosaurs.

Yet in spite of these continuations from the late Mesozoic into the early Cenozoic of established, yet changing continental relationships, of widespread tropical and subtropical environments, and of the rather modern quality of plant life, a line has been drawn to separate the two eras from each other. This line has been determined in part by the changes in animal life, especially by the shift from Mesozoic reptilian dominance to Cenozoic mammalian dominance. It is, above all, the end of the dinosaurs as well as of certain other large reptiles that marks the end of Mesozoic time; it is the appearance of abundant mammals that marks the beginning of Cenozoic time.

EXTINCTION OF THE DINOSAURS

The reasons for the extinction of the dinosaurs and other large reptiles at the end of Cretaceous history cannot be explored at this place. It is a large subject, and although it is not especially pertinent to our discussion, perhaps some indication of the range of ideas concerning this extinction may be of interest. As was remarked in the last chapter, there are numerous theories to account for the extinction of the dinosaurs, and for every theory there are legitimate objections. A few of them may be briefly outlined as follows.

One of the most popular and widely disseminated theories supposes that the dinosaurs were wiped out by some worldwide natural catastrophe. This theory envisages events on a large and dramatic scale. But if there had been some great catastrophe, why was it so neatly confined to the dinosaurs, the flying reptiles, and the great marine reptiles of late Cretaceous time? One would suppose that various other denizens of the earth would have been caught up by such an event.

Perhaps a variant of this idea is the supposition that the dinosaurs were wiped out by some epidemic. Our modern knowledge of diseases, however, indicates that, generally speaking, they are rather specific in their effects. An epidemic that would have killed off all of the highly varied dinosaurs, as well as certain other reptiles, but which would have spared those close cousins of the dinosaurs, the crocodilians, not to mention numerous other backboned animals, is difficult to visualize.

We know that there were environmental changes that marked the close of Cretaceous history. Is it not possible that the dinosaurs succumbed to such changes? But again, why should the extinctions have been along such strict zoological lines? For example, it is generally thought that there was a gradual cooling of climates at the end of Cretaceous history, as a part of the advent of modern mountain building. And reptiles are very sensitive to temperatures, since they have no mechanism for manufacturing significant body heat (at least this is true for all modern reptiles). We know, however, that many modern reptiles protect themselves against cold winters by burrowing into the earth. It patently was not possible for the giant dinosaurs to have escaped from lowering temperatures by burrowing, but could not some of the smaller ones (and there were small dinosaurs) have sought protection in this manner? Moreover, many crocodilians, large and small, survived the Cretaceous-Cenozoic transition; so why should not some of the dinosaurs have so survived?

Conversely, suggestions have been made that the dinosaurs were wiped out by increased temperatures (although it must be admitted that there is little evidence in the geological record to support the idea of high temperatures at the end of Cretaceous time). Again, as in former cases, why should all of the dinosaurs, large and small, be destroyed by such a phenomenon if it did occur?

It has been suggested that dinosaurs were the victims of competition with the early mammals, which were burgeoning at the end of Cretaceous history. Yet the record would seem to show that the real increase of mammals did not come about until after the dinosaurs had become extinct. As long as the dinosaurs were living, their mammalian contemporaries were very small and of generalized types.

Perhaps, say some, the dinosaurs were decimated and finally destroyed by mammals that preyed upon their eggs. Such predation might have had some effect on the fortunes of dinosaurian persistence, but it seems hardly likely that it could have brought about the complete disappearance of these numerous and varied reptiles. Many modern reptiles—the great sea turtles, for instance—are subjected to intense predation of their nests, yet they continue in spite of this. Moreover, there is the good possibility that at least some dinosaurs may have given birth to their young alive, as do many modern reptiles.

Also, it has been suggested that for some reason the eggs of dinosaurs were not viable at the end of Cretaceous time. This suggestion has been based upon the large number of unhatched eggs found in the late Cretaceous sediments of southern France. Yet it must be remembered that the eggs so discovered are probably a minute fraction of all of the dinosaur eggs that were deposited in that region. They may represent the failures; the bones of dinosaurs in the same deposits represent the successes.

Were the dinosaurs the victims of extraterrestrial radiation? If so, why were not other animals also affected? This argument is subject to the same objections as have been made with regard to natural catastrophes.

A suggestion also has been made recently that perhaps the extinction of the dinosaurs was brought about by a change in the ratio of DNA to the cell nucleus in these ancient reptiles. It is an interesting idea, but as yet has no hard evidence to back it up.

When all is said, one comes down to the simple fact that the earth changed, and the dinosaurs were unable to adapt themselves to the changing earth. Which is a nice generalization that does not really say very much. Suffice it to say that at the very end of the Cretaceous the last of the dinosaurs became extinct—this in spite of their history of a hundred million years and more of successful evolutionary development across the face of the earth. And joining the dinosaurs in their geologically rapid descent into oblivion were the flying reptiles, or pterosaurs, as well as the marine reptiles, the ichthyosaurs and the plesiosaurs. Various turtles disappeared, as did some of the crocodilians, but their evolutionary lines were continued, to thrive into modern times. The important fact is that with the close of Cretaceous history the large dominant animals of land and sea disappeared, leaving vacant spaces to be occupied by new dominant animals.

Perhaps the extensive extinctions at the end of Cretaceous time are to be correlated in some way with the beginning of worldwide mountain uplifts, which eventually were to result in the modern great mountain systems—the Himalayas and the Alps, the Andes and the Rocky Mountains. This is known as the "Laramide revolution" in earth history. However that may be, the empirical evidence shows that with the advent of Cenozoic history the mammals inherited the earth.

EARLY RADIATION OF THE MAMMALS

At this place it may be well to review briefly the varied mammals that fell heir to the earth from their reptilian predecessors. Such a review is in a sense another digression from our main interest, but in order to discuss and appreciate the distributions of land-living vertebrates in relation to Cenozoic continental connections the subject perforce is here introduced.

Of course, mammals lived on the Mesozoic continents from late Triassic time throughout the remainder of the era, and they were probably very numerous. But they were small and seemingly inconsequential as long as the dinosaurs remained the dominant land-living animals. Nevertheless, by late Cretaceous time primitive marsupials, some of them not unlike the modern North American opossum, and placental insectivores similar to modern shrews and hedgehogs ranged widely. (As previously noted, it was from an insectivore stem that the tremendous radiation of Cenozoic placental mammals had its origins.) Also there were some Cretaceous holdovers of primitive Mesozoic mammals, of which the multituberculates should be mentioned. These were rodentlike mammals, briefly described in the last chapter, that continued into Cenozoic time, until they seemingly were displaced by the true rodents.

The beginning of mammalian hegemony was marked by an early Cenozoic radiation establishing many interesting fur-clad animals in the ecological niches that had been vacated by the dinosaurs and other

Mesozoic reptiles. The small marsupials and insectivores that had flourished beneath the feet of the giant dinosaurs during Cretaceous times continued and expanded in variety during the inaugural years of the Cenozoic era. Certain other small mammals arose, perhaps not at the very beginning of Cenozoic time, but certainly during the very early phases of Cenozoic history, to add variety to the inhabitants of post-Cretaceous continents. Such, for example, were the bats, undoubted offshoots from the insectivores. The earliest known bats are full-fledged bats, and it seems evident that these small mammals quickly evolved to share in part some of the ecological niches that had been vacated by the flying reptiles. Also primitive carnivorous mammals, known as "creodonts," appeared at an early stage in Cenozoic history to assume the roles abandoned by various predatory reptiles. Strangely enough, creodonts did not appear in South America; there the function of predation was taken over by carnivorous marsupials, known as "borhyaenids." There were other small and medium-sized mammals inhabiting the early Cenozoic continents, such as the very primitive primates (like the bats, direct branches from an insectivore stock), ancestral rodents and rabbits, and the progenitors of the edentates of later times—the anteaters, sloths, and armadillos.

But a particularly distinctive feature of early Cenozoic mammalian faunas was the rapid evolution of large plant-eating mammals, which, it would seem, occupied the niches that had been left vacant by the extinction of the large plant-eating dinosaurs. Among these were the condylarths, which were rather generalized herbivores, in some respects not so very far removed from some of the primitive carnivores, and typified by their short, five-toed feet, rather heavy limbs, long tail, low skull, by the comparatively primitive, low-crowned cheek teeth and by the prominent canine teeth, and from which some of the more advanced plant-eating mammals evolved. Sharing the early continental environments in North America with the condylarths were the pantodonts and the uintatheres—the former being heavily built, short-footed herbivores, the latter being gigantic mammals, characterized by long skulls from the top of which there often arose three pairs of bony excrescences, which might be called horns. Both pantodonts and uintatheres had long, saberlike canine teeth.

At this stage of geologic history there were large herbivores known as astrapotheres and pyrotheres inhabiting South America, the former descended from condylarth ancestors. It would appear that these strange mammals had short trunks, and the teeth in the pyrotheres are so very like those of some of the proboscideans (the group to which the modern elephants belong) that certain early students of South American fossil mammals thought the pyrotheres to be a branch from the proboscideans. We now know otherwise.

In Egypt there have been found the skeletons of arsinoitheres—gigantic, vaguely rhinoceroslike herbivores (quite unrelated to the rhinoceroses) characterized by a pair of immense horns, side by side, protruding up and forward from the top of the skull.

All of these various large herbivores, as well as the early carnivores, lived successfully through many millions of years of early Cenozoic history. But eventually they became extinct, to be replaced by the herbivores and carnivores that inhabit the world of today. The early herbivores were characterized by low-crowned cheek teeth, which were quite adequate for the soft vegetation of ancient Cenozoic landscapes. But when hard grasses developed these animals were not adapted for such a diet, so they gave way to the "modern" herbivores, in which the cheek teeth have tall crowns and can thus withstand the wear resulting from a persistent diet of abrasive plants. Furthermore, these early plant eaters were in a sense "clumsy." Certainly they were not fleet of foot, which again put them at a disadvantage when compared with the more modern, long-legged herbivores. The primitive carnivores, adapted for preying upon such relatively slow plant-eating mammals, disappeared when their prey became extinct. The world of later Cenozoic times was to be inhabited by very agile beasts of prey, capable of pursuing fleet browsers and grazers. Finally, these early Cenozoic mammals, herbivores and carnivores, were characterized by small, primitive brains, ill-suited to a world in which intelligence was becoming ever more important in the struggle for survival.

THE EVOLUTION OF MODERN MAMMALS

Consequently the early Cenozoic mammals became extinct, their places being taken by the herbivores and carnivores of later Cenozoic and recent time. One group of these herbivores is that of the perissodactyls, or odd-toed ungulates, in which there commonly are three functional toes or a single toe in each foot, and consisting of horses and their relatives, rhinoceroses and tapirs, as well as certain groups now extinct, notably the gigantic titanotheres, many with rather bizarre "horns" or bony excrescences on the skull, and the strange, clawed chalicotheres, somewhat horselike in appearance, in spite of their clawed

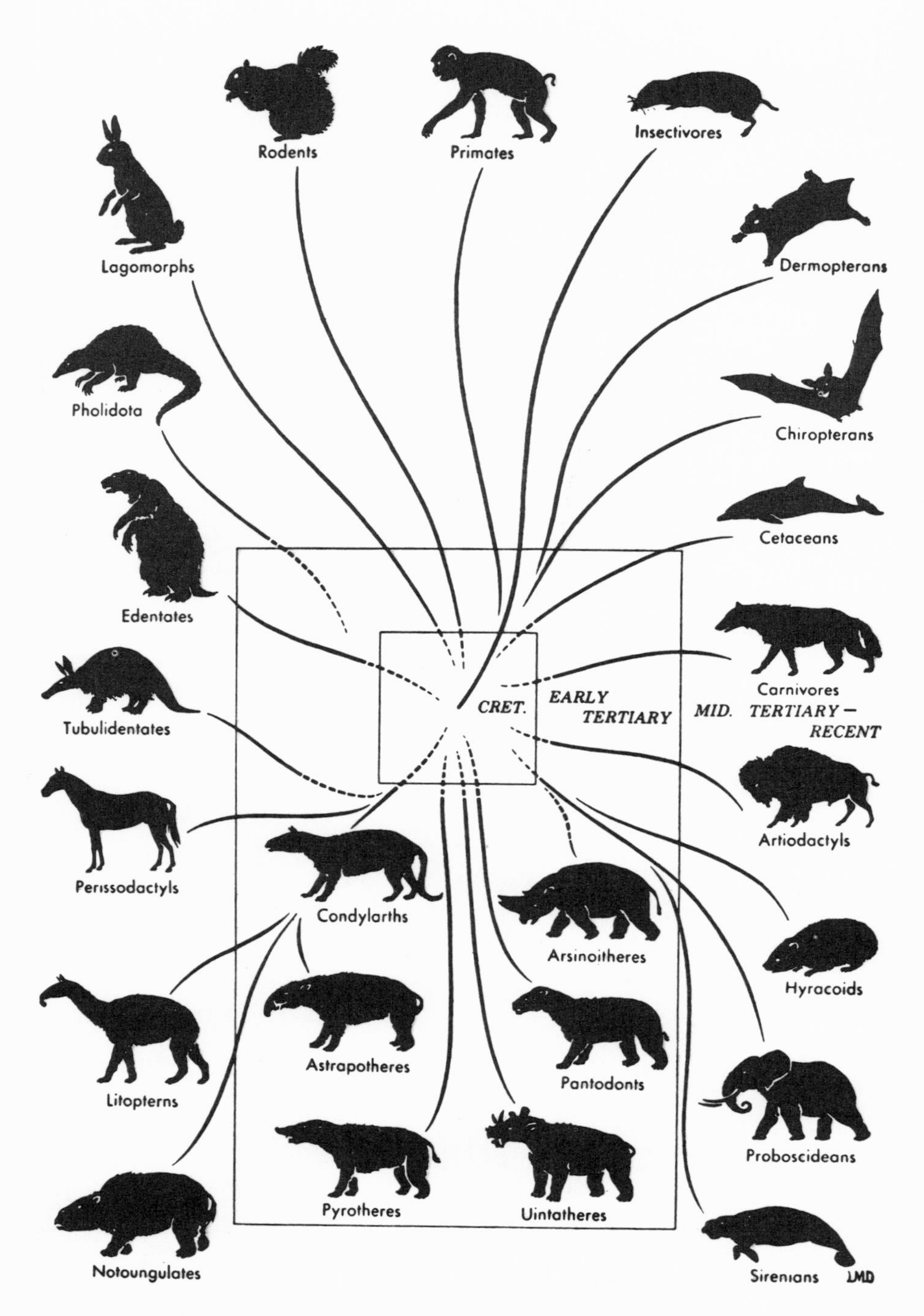

FIGURE 79. The radiation of placental mammals from a Cretaceous ancestry centered within the insectivores. Certain mammalian orders were restricted to early Tertiary times. The majority of orders, except for the litopterns and notoungulates (which became extinct during the Pleistocene epoch), have survived into modern times.

feet. Another herbivorous order is the much more varied artiodactyls, or even-toed ungulates, with two or four toes in each foot; the pigs and peccaries, hippopotamuses, camels and llamas, deer, giraffes, the pronghorns of North America, the numerous antelopes, sheep, goats, and cattle, and several groups now extinct, of which may be mentioned the gigantic, piglike entelodonts, often as large as modern bison, and the rather isolated oreodonts (confined to North America), not easily compared to any modern mammals, but in some respects like sheep, to which they were not in the least related, aside from being artiodactyls. Also there is the group of proboscideans, already mentioned, eventually evolving as the giants among land mammals, containing the extinct moeritheres, mastodonts and dinotheres, and the elephants, two genera of which inhabit our modern world. Related to the proboscideans are the hyracoids, represented in the world today by the conies or dassies, living in Africa and the Middle East; and by the aquatic sirenians, or sea cows.

In South America the dominant herbivores of later Cenozoic time were the varied notoungulates, some light and agile, others heavy and as large as rhinoceroses, and the rather camellike and horselike litopterns. These numerous plant eaters were, like the astrapotheres and pyrotheres previously mentioned, peculiar to the South American continent. Sooner or later they all became extinct.

As for carnivorous mammals, the ancient, relatively clumsy creodonts were replaced by the carnivores that we know—raccoons, bears, wolves and dogs, weasels, otters, mink and marten, civets, hyenas, and the cats, large and small. Certain carnivores have become adapted for life in the oceans, these being the sea lions, seals, and walruses.

There might be mentioned in passing the mammals supremely adapted for salt-water life, the cetaceans, or whales and porpoises.

While the placental mammals were evolving in these several directions, the marsupials continued and in some regions, notably South America and Australia, became quite various. Australia today is the home of kangaroos and wallabies—the large herbivores of that continent—of wombats, phalangers, and a host of other pouched mammals, many of small, mouselike aspect; and until the subrecent past, of a very active, wolflike carnivore, the thylacine, generally believed extinct, but possibly still living in the remote parts of western Tasmania. In Australia, too, are the egg-laying monotremes, the echidna and the platypus, representing a completely separate line of mammalian evolution for which there is virtually no fossil record.

The carnivorous marsupials of South America, living through many millions of years of Cenozoic history, have already been mentioned. A few small groups of mammals, the extinct taeniodonts and tillodonts, the extinct tusked, mastodontlike desmostylids—related to the sea cows—and the modern colugos, or "flying lemurs," as well as the aardvark of Africa and the Oriental scaly pangolin, may be added to this mammalian roster.

A SUMMATION OF MAMMALIAN EVOLUTION

Such was the radiation of the vertebrates that dominated the continents of Cenozoic time. And this development of Cenozoic mammals was a complex history with two distinct phases. The first phase was, as we have seen, an early Cenozic one, in which many of the modern orders of mammals such as insectivores, bats, primates, rodents, rabbits, edentates, and some others became well established. And it was also a phase when large primitive herbivores roamed widely across the continents, to be preyed upon by primitive carnivores, the so-called creodonts, and in South America, the carnivorous marsupials. The second phase was inaugurated by the extinction of the primitive herbivores and carnivores as a result of competition from more advanced plant-eating and meat-eating mammals. During this second phase many of the orders of mammals that had become so well established in early Cenozoic history continued uninterruptedly their evolutionary development; such were the insectivores and other mammals just listed. And it should be remarked that during this later phase of Cenozoic history the advanced primates made their appearance, some of them eventually evolving as the great apes and as men. The significance of this aspect of mammalian evolution need not here be labored.

The first phase of Cenozoic mammalian evolution took place during the Paleocene and Eocene epochs, the first two subdivisions of the Tertiary period of earth history. The second phase had its proper beginnings in the Oligocene epoch, the third of the Tertiary subdivisions, and continued through the last two of the Tertiary epochs, the Miocene and Pliocene, and on into the Pleistocene epoch (the great Ice Age) of the Quaternary period. Some of the primitive mammals continued into the Oligocene epoch and even later, but by and large the Oligocene can be regarded as a time when mammalian faunas began to assume their modern aspects.

Such is a brief outline of the mammalian groups that evolved and dominated the Cenozoic continents. It may be added that during this great division of earth history the birds radiated along multitudinous lines as we know them, while the reptiles left over from the Mesozoic—the turtles and crocodilians, the lizards and snakes, and the lone tuatara of New Zealand—continued quite successfully. Amphibians were represented by the members of this group, which still inhabit the earth—the frogs and toads, the salamanders, and the tropical apoda.

MAMMALS OF THE ANCIENT CENOZOIC CONTINENTS

With these necessary and perhaps boring details set forth, we may now turn our attention to the problem of Cenozoic tetrapod faunas and continents.

The record of Cenozoic mammals in North America is especially full and rich—more completely documented than in any other continental region. It is a record beginning almost immediately after the cessation of dinosaurian burials; in Montana, for instance, the first primitive mammal faunas are separated from the last of the buried dinosaurs by only a few feet of clays and sandstones—sediments that continue without change from the Age of Dinosaurs into the Age of Mammals. As we have noted, the beginning of the Cenozoic world was very much like the end of the Mesozoic world, and here in Montana evidence is proof of the statement. Whatever the physical conditions and environments may have been in this western portion of the North American continent, they obviously continued without interruption during that important transition from dinosaurian dominance and extinction to mammalian rule and increase.

From such a beginning the mammals evolved and increased within North America through successive faunas that record the changing quality and quantity of tetrapod life on a geologically evolving continent. The earliest of the Paleocene faunas contained small primitive mammals: opossumlike marsupials and tiny insectivores, rodentlike multituberculates that preceded the true rodents in the economy of ancient mammalian life, very primitive creodonts—the most ancient of the carnivorous placental mammals and not very far removed in structure from their insectivore ancestors—some small ancestral condylarths, which more or less fulfilled the function of hoofed mammals in those days, and certain strange mammals peculiar to early Cenozoic times, such as the comparatively large taeniodonts.

As the Paleocene epoch drew to a close other mammals were

added to this roster of early furred animals, to lend variety to the tetrapod faunas of those days. Three or four developments are of particular interest. It was at this time that early primates appeared—very primitive primates, indeed—which shared many of the characters of their insectivore ancestors. Nevertheless they marked the beginning of what seems to us the most important line of mammalian evolution. Also there appeared some small mammals known as "metacheiromyids," which would seem to have approximated the ancestors of the edentates—the mammals that were to loom so large in the development of South American faunas. But of especial interest was the advent in late Paleocene time of enlarged, and one might say "improved," carnivores and herbivores. Some of the creodonts of this time were as large as modern wolves, some of the condylarths became as large as modern sheep, and there entered on the scene the first uintatheres, the giants of early Cenozoic history.

This long fossil record, extending through the ten-million-year duration of the Paleocene epoch and demonstrating the origins and the evolution of early mammalian faunas, is unique; in no other part of the world is it duplicated. When we look to the west, to Asia, we see late Paleocene mammals in the Gashato Formation of Mongolia; when we look to the east, to Europe, we see equally late Paleocene mammals in the Cernay beds of France and in the somewhat older Thanet sands of England; and when we look to the south, we see late Paleocene mammals in the Rio Chico beds of Argentina. What are the relationships of the late Paleocene faunas in these Asiatic, European, and South American regions to those of western North America?

The Mongolian Gashato fauna is made up of marsupials, multituberculates, insectivores, primitive creodonts, and various early herbivores, such as are characteristic of late Paleocene time. All of these early mammals are closely comparable and related to their counterparts in the late Paleocene of North America. Also, an interesting and unexpected animal in the Gashato fauna is *Palaeostylops,* a primitive notoungulate closely comparable to another northern notoungulate, *Arctostylops,* which occurs in the Upper Paleocene sediments of Wyoming. Both of these animals, although small and of primitive herbivore aspect, show characters prophetic of and related to the features that so characterized the most numerous plant-eating mammals of Cenozoic South America. It would seem evident that there was close communication between northeastern Asia and western North America during this final phase of Paleocene history, a relationship that shortly was to be extended into South America. A very recent paper, however, advances a different interpretation to explain the origins of the South American mammal faunas. (Fooden, 1972, pp. 894–898.) According to this theory, South America was an island continent by Cretaceous time, when it already was inhabited by marsupials and primitive placental mammals. These mammals presumably were isolated from related mammals in other continents before the advent of the Cenozoic; therefore the evolution of the marsupials, edentates, condylarths and notoungulates would necessarily have proceeded separately and along lines parallel to such of these mammals that might have lived outside of South America during early Cenozoic time. This explanation assumes an earlier presence of placental mammals in South America than is indicated by the known paleontological record.

In the Upper Paleocene beds of Europe there occur multituberculates, condylarths, and creodonts of types that are found in North America, but not in Mongolia. Furthermore, primitive primates also occur in Europe as well as in North America, but are lacking in the Mongolian sediments. Evidently there was communication between North America and northern Europe in late Paleocene time, but obviously by a path different from that which connected North America and northeastern Asia.

It thus seems established that at this stage of earth history the trans-Bering land bridge was functional, as it had been during the days of the last dinosaurs, to join eastern Asia and North America, while at the same time there was a North Atlantic bridge, again as there had been during the final years of dinosaurian dominance, to join western Europe and North America. The connections of North America with the east and the west, but the lack of identity between the European and Asiatic Paleocene faunas, can mean only one thing—that there must have been a barrier between the two regions of what is now a single continental land mass. And that barrier probably was the north-to-south seaway in the general region of the present Ural Mountains, a seaway that had persisted from Cretaceous time. Laurasia of late Paleocene time was thus subdivided, as it had been during late Cretaceous time, by the marine transgression extending northward from the ancient Tethys Sea. The subdivision was, of course, superficial in geological terms—a shallow and narrow flood across the Laurasian land mass. Nevertheless it was quite sufficient to prevent the transverse movements of primitive mammals back and forth between the European and eastern Asiatic regions. Western Europe and eastern

Asia were, in effect, the outliers of a central North American land, a vast area where Paleocene mammals flourished in abundance and variety.

The occurrences of Upper Paleocene notoungulates in Mongolia, and again in North America, are significant, because they indicate a long line of overland communication from eastern Asia through western North America and on into the southernmost part of South America, in which last region various late Upper Paleocene notoungulates are to be found. Quite obviously these animals moved between the three continental regions by land bridges, and quite obviously these bridges were the trans-Bering connection and an isthmian link in the region of the present Panamanian isthmus, unless one accepts the independent evolution of these animals in South America since Cretaceous time. Other early South American mammals, especially condylarths, support this evidence.

PALEOCENE CONTINENTAL CONNECTIONS

Such were the continents and their tetrapod inhabitants of Paleocene time, when mammals had extended their sway over lands that had for so long been ruled by the dinosaurs. Laurasia was a divided land. South America, by now completely separated from Africa, was perhaps connected by an isthmian link with North America. And the early mammalian faunas were extended across these lands, spreading from North America to Asia by a trans-Bering bridge, from North America to Europe by a North Atlantic bridge, and seemingly from North America to South America by a Panamanian isthmus. As for those parts of Gondwanaland other than South America—Africa, peninsular India, Australia, and Antarctica—fossil evidence of land-living vertebrates is as yet to be found. Geological facts would indicate that there was a connection from South America into Antarctica by way of the Antarctic peninsula, and that Antarctica and Australia still were connected. Perhaps it was through these land connections that the marsupials spread between South America and Australia. Perhaps future discoveries in Antarctica will supply the now missing fossils that will provide a key as to the origins of the unique marsupial faunas of Australia.

Africa may still have been joined with lands to the north, or it may have been during this span of geologic time a great island continent. There are no records of Paleocene tetrapods to give us the clues that would be so helpful. And the same is true for peninsular India. Might it be that at this time peninsular India was the island continent so often envisaged, moving in isolation, or perhaps accompanied by China, across the Indian Ocean from its original position as a part of Gondwanaland to its future position as a part of Asia?

THE NORTHERN HEMISPHERE MIGRATIONS OF EOCENE MAMMALS

The close connections between North America and northeastern Asia, on the one hand, and northwestern Europe, on the other, continued from Paleocene into Eocene times. Again, as in the Paleocene, the early Eocene connections between North America, here regarded as a centrally placed land mass, with the Asiatic and European land masses on either side of it, were essentially almost mutually exclusive. That is to say, the land-living mammals that moved back and forth between North America and northeastern Asia did not reach Europe, while those that made the crossings between Europe and America did not extend their ranges beyond into Asia.

Although the obvious relationships between Lower Eocene mammals in North America and northeastern Asia are strong, those between the mammals of North America and Europe are truly remarkable. It has

FIGURE 80. The distribution of the ancestral horse *Hyracotherium* (*Eohippus*) during Eocene times. The existence of a Laurasian continent, with Europe and North America connected, explains not only the presence of *Hyracotherium* in the two regions but also the remarkable similarity of Eocene mammalian faunas of North America and Europe.

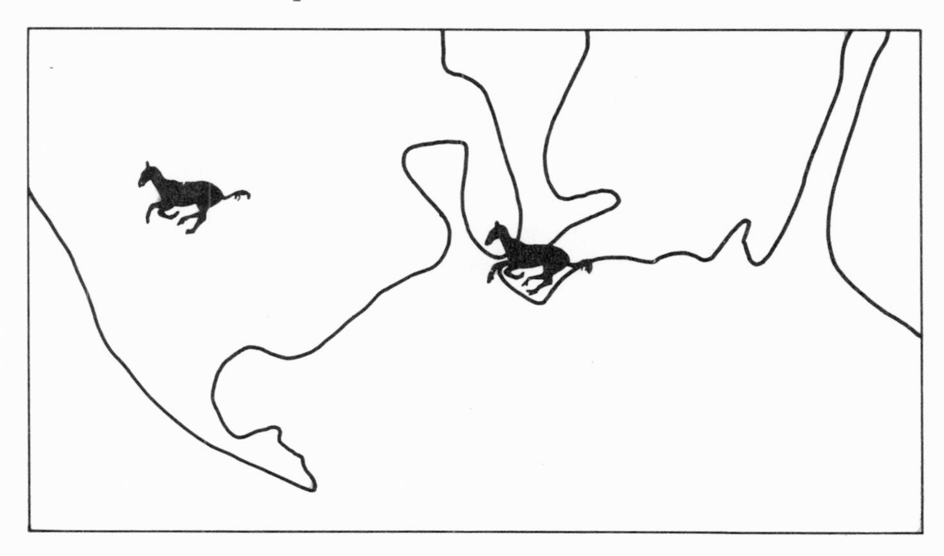

been said that at this stage of earth history the faunal resemblances between North America and Europe are such as to indicate these two areas to have been zoogeographically a single region. There are numerous identities in genera, notably among the primitive carnivores or creodonts, the rodents, certain primitive herbivores, and, of particular interest, the primitive horse, *Hyracotherium*—very commonly referred to in the literature as *Eohippus*. The considerable number of identical genera (and probably in some cases, at least, of identical species) inhabiting Europe and North America can only mean that there was a broad and intimate connection between these regions, a connection that certainly made of them a single land surface across which was distributed a fauna of continental proportions.

In the early horse, *Hyracotherium*, we see among the varied archaic mammals—creodonts and ancient herbivores—the influx of one genus representing a modern group of herbivorous mammals. It was a small beginning, but was prophetic of the revolution in mammalian life which, beginning at about this stage in the history of the earth, would by the close of Oligocene times effectively replace the ancient and archaic groups of mammals with those groups of mammals, especially the carnivores and herbivores, which are today dominant.

To return to northeastern Asia, strong resemblances of its Eocene mammals to those of North America are seen among the creodonts, certain primitive herbivores, particularly uintatheres, and, representing the modern herbivores, early tapirs. From this it is evident that the trans-Bering connection was still effective.

These geographical and faunal relationships between different regions of what was then a single great transverse continent were destined to change with the progression from early Eocene through middle and late Eocene history. It would appear that there was little movement of animals between Asia and North America during middle Eocene time; perhaps there was a shallow oceanic barrier between the two regions, as there is today. However, the interchange of faunal elements would seem to have been resumed with the beginning of late Eocene history, to continue from then until the end of the epoch. The late Eocene movements of mammals across the trans-Bering region were dominated by early perissodactyls, the odd-toed hoofed mammals that even at this stage of earth history were beginning to displace the more archaic herbivores that heretofore had inhabited the northern continents. These invading perissodactyls were titanotheres (large and even gigantic browsing plant eaters that fulfilled the roles in the ecology

FIGURE 81. An early Eocene landscape in North America. In the left foreground is the archaic carnivore (or creodont) *Oxyaena,* facing the small, ancestral horse *Hyracotherium* (frequently designated as *Eohippus*). Behind the two horses is *Phenacodus,* an early, generalized hoofed mammal belonging to the order *Condylarthra.* Behind *Phenacodus* are other archaic hoofed mammals, *Coryphodon,* representing the order Pantodonta.

of Eocene and Oligocene lands that subsequently were to be assumed by the rhinoceroses), varied tapirs, and even a few early rhinoceroses. Also there were very primitive chalicotheres, these being the anachronistic perissodactyls previously mentioned, in which the toes were armed with large claws rather than with the usual hoofs that one expects in plant-eating animals. And in addition there were a few artiodactyls, the even-toed hoofed mammals that today are at the zenith of their evolutionary development. The representatives common to Asia and North America at that time were large piglike entelodonts (which, it must be emphasized, were not pigs) and small antelopelike mammals, known as "hypertragulids" (which likewise must be excluded from the true antelopes). We see here certain manifestations of artiodactyl evolution that preceded and eventually were replaced by the artiodactyls (e.g., pigs and antelopes) we are familiar with today.

Accompanying these multitudinous early perissodactyls and considerably more restricted early artiodactyls in their movements between Asia and North America were various archaic creodont carnivores, as well as representatives of modern carnivorous groups such as the cats and canids, and in addition insectivores, rodents, and rabbits.

It is evident that the mammals wandering back and forth between Asia and North America in late Eocene times, although retaining certain primitive representatives, were on the whole assuming the appearance of modernity. Cats and dogs (using these terms in a broad sense), rodents and rabbits, tapirs and rhinoceroses, even though in the early stages of their evolutionary progress as the Eocene epoch was coming to a close, nevertheless have about them a familiar look. It was the beginning of the development of modern faunas—a development that was to proceed from then until recent times. In a more restricted sense it was the beginning of related developments among North American and Asiatic mammalian faunas, which, owing to the intermittent presence of a trans-Bering connection, were to mark the history of land-living mammals in a great latitudinal circle around the north polar regions during millions of ensuing years. We will see the continuation of this story when we review the relationships of continents and land-living mammals of Oligocene time.

The movements of mammals between North America and Asia across the trans-Bering bridge was to be a continuing story; the movements between North America and Europe across a North Atlantic ligation was to come to an end. And this ending was an event probably of middle and late Eocene history. The drift of the continents proceeded, and it is likely that there was a beginning of the rift in Laurasia that eventually was to establish Europe and North America as separate lands, with Greenland (by definition a part of North America) occupying a position intermediate between the two continents. Certainly from about the end of Eocene time there was no longer the direct interchange of mammals between northwestern Europe and North America that so prominently characterized faunal relationships of early Eocene times. From now on any interchange of mammals and other land-living vertebrates between Europe and North America would be by another and more circuitous route.

At the same time, we begin to see the beginnings of intermigrations between Europe and Asia to the east. So it would seem that the narrow incursion of marine waters northward across Asia from the Tethys Sea had begun its retreat. In the future there was to be a single Eurasiatic land mass, allowing for movements of land-living animals back and

FIGURE 82. Middle Eocene North America. The large herbivore with six horns on the head is *Uintatherium,* representative of the order Dinocerata. These mammals, conveniently called the uintatheres, were gigantic plant eaters, ecologically replaced in later Tertiary epochs by the titanotheres and the rhinoceroses. At the base of the tree is the creodont (or archaic carnivore) *Patriofelis.* A modern touch is added to the landscape by the squirrellike rodents *Paramys* and the early lemur *Notharctus,* in the tree.

forth, east and west. There was an end to the distinction of western European from eastern Asiatic land-living vertebrate faunas, which had held since Cretaceous times, since the days of the later dinosaurs. Ancient Laurasia was coming to an end; Holarctica, the great northern circumpolar land mass, was coming into being.

THE ISOLATION OF SOUTH AMERICA, AUSTRALIA, AND ANTARCTICA

We have seen that probably there was an isthmian link between North and South America during Paleocene time—a link that enabled condylarths and condylarth-derived ancestors of the strange isolated South American herbivores to become established on the southern continent. In particular, as we have seen, the presence of very archaic notoungulates, *Palaeostylops* in Asia and *Arctostylops* in North America, both closely related to the various primitive notoungulates found in South America, give added proof as to the presence of an isthmian link at the beginning of Cenozoic history. It is evident, however, that at some time during the Eocene epoch this link was broken, not to be reestablished until well into Pliocene times. Thus South America became essentially an island continent and remained a secluded land for many millions of years, as is abundantly shown by the succession of mammalian faunas quite unlike those found in other parts of the world. The mammals of South America, living in isolation during most of the span of Cenozoic history, were to evolve along lines of their own, as will subsequently be made evident.

Geological evidence indicates that Australia and Antarctica, which had retained their ancient connection as constituent parts of Gondwanaland through Paleocene history, probably became separated from each other during Eocene times. Thus Australia began its long separation as an island continent—an isolation that continued from that day unto this. The strange mammalian faunas unique to South America were in part suppressed and in part leavened by the influx of northern mammals, when connections were reestablished between that southern continent and North America during the Pliocene epoch, but the equally strange mammalian faunas of Australia remained intact and inviolate until that continent was invaded by eighteenth-century Europeans. More of this later.

As for Antarctica, the Gondwana counterpart of Australia (these being perhaps the last remnants of Gondwanaland to be separated from one another), there is nothing to be said. Our knowledge of Cenozoic land-living vertebrates on the south polar continent remains virtually a blank. It may be that future explorations, especially in the Antarctic peninsula, will reveal something of such vertebrates on that now ice-locked island continent.

WHERE WAS CENOZOIC INDIA?

Peninsular India likewise is a blank, so far as Paleocene to Pliocene land-living vertebrates are concerned. The evidence is not there, or has not been found. Could this be the time, as already suggested, when peninsular India, detached from its original position as a part of Gondwanaland, was moving northeastward toward its eventual junction with Asia? This is a possibility to be kept in mind. It will be most interesting, if Lower or Middle Cenozoic land-living mammals are found in the Indian peninsula, to see what they will show.

In this connection one might note that Eocene mammals have been found in Burma. These, however, are outside the limits of peninsular India, and thus have nothing to do with the history of that subcontinental mass. The Burmese fossils, among which are early primates, titanotheres, tapirs, rhinoceroses, and artiodactyls known as "anthracotheres," form part of the story of mammalian evolution in Asia.

THE EARLY MAMMALS OF AFRICA

It was noted that vertebrate evidence concerning Africa in Paleocene time is lacking. The same is true for much of the Eocene. But in sediments of late Eocene and early Oligocene age, designated as the Qasr-el-Sagha and Jebel-el-Quatrani formations respectively, exposed in the Fayûm region of Egypt, there have been discovered through the years treasure-troves of ancient and often unique mammals. Here, along the northern border of the African continent, is revealed what appears to be the locus of origin for certain mammals not found elsewhere in the early Cenozoic world. In brief, we see in the Fayûm beds ancestral proboscideans—the order of mammals represented today by the elephants—ancestral sirenians, or sea cows, and ancestral hyracoids—the mammals represented in our modern world by the biblical conies of the Middle East and the dassies of Africa. However, only the comparatively small and trunkless moeritheres, the very first of the proboscideans, and a few other scattered mammals are found in the Eocene levels of the Fayûm sequence. We will look at the mammals of the succeeding Lower Oligocene portions of the Fayûm beds in the next chapter, where the continental relationships of Oligocene land-living vertebrates are considered.

It should be mentioned here that, although the history of Cenozoic land-living vertebrates is virtually absent in Africa until we reach the Upper Eocene–Oligocene Fayûm beds, there are evidences of Paleocene and Eocene mammals, as well as those of later Cenozoic epochs, in Madagascar. In the Paleocene of Madagascar are found fossil insectivores; in the Eocene, fossil lemurs. Projecting our story a bit, fossil fossas (which are catlike civets) occur in the Oligocene, rodents in the Miocene, and a hippopotamus in the Pleistocene. The limited and unbalanced nature of these fossil discoveries indicates quite strongly that very probably by the beginning of Cenozoic time Madagascar had become separated from the African mainland and during the subsequent years of the Cenozoic received such mammals as lived there by adventitious circumstances, probably by the accidental and very occasional arrivals of colonizers on floating logs or masses of vegetation.

This brings us to the end of the story of early Cenozoic, or Paleocene–Eocene, continents and mammals. It is the end of the prelude to the modern world. It is the end of domination of the lands by archaic mammals—mammals that by the close of the Eocene epoch were being supplanted by the progressive mammalian lines that have survived into our modern world. We now turn our attention to a world in which continents and mammals were indeed assuming the positions, the relationships, and the characters by which we know them.

CHANGING PATTERNS OF DISTRIBUTION

In taking our leave of the ancient Cenozoic world, we are leaving a globe on which the continents, still drifting, were making the final breaks that disrupted the old Laurasian and Gondwanaland masses. Indeed, those ancient supercontinents were by now hardly recognizable. South America had broken away from Africa, as had Antarctica and Australia, while these latter two continents became separated from each other probably during the Eocene epoch. India seemingly was on the move from its original position, toward Asia. And Laurasia, although essentially a single mass with trans-Bering and North Atlantic ligations, was for at least a part of this interval nevertheless bisected by a narrow, shallow sea, flooding northward across Asia from the Mediterranean Tethys.

These various relationships between the continents were reflected in the distributions and relationships of mammalian faunas. The old Gondwanaland relationships had virtually vanished. Land-living vertebrates were now becoming established in patterns leading to the faunal patterns of today. There were differences, it is true, because this was early in the history of mammalian evolution and in the development of mammalian faunas. Yet on the whole the modern system of continental positions and connections and the resultant arrangements of mammalian faunal relationships were becoming evident. It is now our purpose to trace the story of wandering lands and animals through its final phases.

IX. MAMMALS AND BRIDGES

CONTINENTS AND BRIDGES

The world of middle and later Cenozoic time, that is, from the beginning of the Oligocene epoch through much of the Pliocene, was a world in part of continents joined by bridges and in part of isolated land masses. A dominant fact in this world was the development of land bridges, during some portion of the Oligocene and much of the Miocene and Pliocene epochs, to unite North America, Eurasia and Africa, thus forming a vast connected land mass surrounding the north polar region of the globe and extending below the equator in that portion of the eastern hemisphere immediately to the east of the present zero meridian. And a dominant fact with regard to this North American–Eurasian–African land mass was the long persistence (with occasional interruptions) of the trans-Bering land bridge. Indeed, this essential relationship of the northern-hemisphere continents and Africa still persists, but with the presence of barriers—the shallow Bering Strait and the immense Himalayan range—that now prevent what was once a considerable interflow of land-living animals.

Contrasted with the ligation of the northern-hemisphere continents and Africa, so characteristic of much of Cenozoic history, was the isolation of continents in the southern hemisphere. Of course, Africa was not isolated, for it had its communications with the lands to the north. But South America was essentially an isolated continent during Oligocene and Miocene time and well into the Pliocene epoch. Australia, probably breaking away from Antarctica during the Eocene, had achieved by the beginning of the Oligocene the status of an island continent, which it has retained ever since. Antarctica, too, was an isolated continent, even though there may have been some connection to the tip of South America, by way of the Scotia Arc. And although there is still much to be learned as to the position of peninsular India during this span of geologic time, it seems likely that it, too, may have been an island mass, moving in the direction of the greater Asiatic continent.

These several continental relationships, as inferred from the evidence of geology and paleontology, had profound effects on the distributions of Oligocene, Miocene, and Pliocene land-living animals and especially on the mammals. Consequently the history of mammalian life and distribution in North America, Eurasia, and Africa was quite distinct from that of mammalian life in South America and Australia, as we shall see. As to the history of mammals in Antarctica (if indeed mammals did live on that now icebound continent during the geological epochs with which we are now concerned) and in peninsular India, the record of the rocks has as yet yielded no definitive information. We must wait to see what future explorations will reveal—if they reveal anything. Let us now turn our attention to the distributions of mammalian faunas in North America, Eurasia, and Africa during the later Tertiary time span that encompasses the Oligocene, Miocene, and Pliocene epochs.

HOLARCTICA

It will be recalled (from the last chapter) that the narrow but elongated north and south seaway that had extended up from the ancient Tethys Sea to divide Eurasia into eastern and western moieties during at least later Cretaceous time, and during Paleocene and much of Eocene time as well, seemingly retreated from the surface of the Eurasian land mass at about the close of Eocene history. Thus Eurasia became a single continent, joined throughout much of its Cenozoic history by the trans-Bering bridge to North America. Thus a united Holarctica, a land in which there was generally a broad link across the

Bering region and in which the Himalayan barrier was comparatively low, became a determining factor in the development of northern-hemisphere mammalian faunas. And thus Africa became in a sense an immense adjunct of this circumpolar Holarctic land mass.

Although with the advent of the Oligocene epoch the unity of Holarctica was established, there was an interruption, within this time span, of the trans-Bering bridge, which had united the North American–Asian regions through long millennia of preceding earth history and which was again to unite these regions during subsequent geologic epochs. Therefore for a time those lands which today are designated as the Palearctic and Nearctic zoogeographical regions, the former being composed of Eurasia north of the Himalayas, the latter of North America including a portion of northern Mexico, were separated from each other, with a consequent effect on their developing mammalian faunas.

OLIGOCENE HOLARCTIC FAUNAS

Perhaps the classic Oligocene faunas of North America are those found in the sequence of deposits that surround the Black Hills of South Dakota—the sediments that collectively are known as the White River beds and individually as the Chadron and Brulé formations, in sequence from bottom to top, or from oldest to youngest. These sediments constitute the famous White River Badlands, from which during the past century there have been collected the abundant remains of land-living vertebrates that inhabited this part of the continent throughout Oligocene history. Similar sediments, containing similar animals, are found in other parts of western North America, to indicate the wide extent of Oligocene faunas on this continent.

Oligocene sediments are exposed at many localities in Europe; particular mention might be made of the upper portion of the Quercy phosphorites in France (the lower Quercy beds being of Eocene age), as well as the French Ronzon beds, all these being of early Oligocene age. The Middle Oligocene is continued by the Quercy sediments, as well as by the Weinheim beds of Germany, while sediments of late Oligocene age are found at various localities in France as well as in the Mainz Basin of Germany.

The Oligocene is represented in Asia by the Lower Oligocene Ulan Gochu and Ardyn Obo formations and the Upper Oligocene Hsanda Gol and Houldjin formations of Mongolia. In Africa are the upper

Fayûm sediments—Lower Oligocene beds that have yielded the bulk of the vertebrates collected from the Fayûm locality.

Such are the regions from which our knowledge of Holarctic and African Oligocene vertebrates, principally mammals, has been obtained. The reconstructed picture of Oligocene faunas as they were distributed throughout this vast continental area, and through the ten-million-year time span of Oligocene history, is a dynamic, changing view that is anything but simple. One would hardly expect it to be otherwise, seeing that we are involved with numerous, abundantly represented faunas, having all of the complexities resulting from the interactions between contemporaneous faunal assemblages, as well as from the successions through time of related antecedent and resultant groups of land-living vertebrates.

Generally speaking, the Oligocene mammals of the Holarctic and African regions have a decidedly modern appearance as contrasted with the Eocene faunas that preceded them. This is owing not so much to the evolution of new kinds of animals (because a large proportion of the modern groups of mammals with which we are familiar had become

FIGURE 83. The Oligocene White River Badlands of South Dakota as they appeared to early explorers. This is the earliest published view, which appeared in 1852, based upon a field sketch made in 1849.

FIGURE 84. The White River Badlands as seen by the modern paleontologist. A skeleton of the creodont *Hyaenodon* is being collected beneath the sunshade on the side of the hill.

established during Eocene times) as to the extinction of varied kinds of archaic animals that had dominated early Cenozoic lands. In terms of human experience, the Oligocene epoch was incredibly distant in time; in terms of mammalian evolution, the outlines of life as we know it were taking definite form.

Of particular interest are the changing relationships of Oligocene faunas in Holarctica. As has been mentioned, the beginning of Oligocene history was marked by the juncture across the Bering region of North America and Eurasia, continuing the bridge that had existed in Eocene time, and this is reflected by the similarities between Lower Oligocene faunas throughout Holarctica. As also has been mentioned, there was evidently a separation of North America and Asia after the early phases of Oligocene history, and this, too, is reflected by the growing dissimilarities between faunas in the two regions.

To begin with early Oligocene history, there are numerous resemblances between animal assemblages found in the Chadron beds of South Dakota, for example, and some of those typical of the Lower Oligocene sediments of Europe, as well as of the Ulan Gochu and Ardyn

Obo beds of Mongolia. Such similarities would seem to reflect the movement of mammals back and forth across the trans-Bering bridge—a continuation of the intercontinental wanderings that had characterized late Cretaceous and early Cenozoic history. It would seem likely that among various mammals that crossed from Eurasia into North America were insectivores, lagomorphs (hares and rabbits), and carnivores of modern aspect, including ancestral canids (the stock from which modern wolves and their relatives developed), progenitors of the raccoons, early mustelids (represented today by the weasels and their kin), saber-toothed cats as well as "typical" felids, and a curious artiodactyl herbivore, a long-snouted mammal in which each foot was provided with four small hooves belonging to the group known as "anthracotheres" and represented in this case by the genus *Bothriodon*. In the other direction, early squirrels, titanotheres, and aquatic rhinoceroses made their way from North America into Eurasia.

But along with these migrants there were numerous "stay-at-homes"—animals that were confined to their own continental regions, many of them being descendants of Eocene ancestors that had lived in these same regions. In Eurasia, in such a category, were moles, various rodents, primitive creodont carnivores, including the genus *Hyaenodon*, viverrids (represented among today's mammals by the Old World civets), perissodactyls known as "palaeotheres," these being rather large, tapirlike odd-toed herbivores, and varied artiodactyls, or even-toed herbivores, among which may be listed the first of the true pigs, the anoplotheres, which were artiodactyls that imitated to a degree the perissodactyls in the structure of their feet in that they had three functional toes instead of the expected four, the cainotheres—small, somewhat harelike artiodactyls—and ancestral deer. In North America there were rodents not found in Eurasia, a primitive carnivore—the creodont *Hyaenodon*, present also in Europe and in both of these regions probably a descendant of Eocene ancestors—early horses, the sheeplike oreodonts and their relatives, the agriochoerids, which evolved in and were confined to North America, primitive camels, and bizarre artiodactyls known as "protoceratids," generally rather like antelopes, but notable by reason of the fact that the skull often was ornately decorated with several pairs of horns on the nose and above the eyes.

In reviewing this list one is struck by the seeming erratic occurrences resulting from animals crossing and recrossing a filter bridge—albeit a broad bridge of the trans-Bering type. Why, for example, did not the active viverrids, the true civets, cross from Asia into North

America? Similarly, why did not the tapirlike palaeotheres make the crossing, as well as various Old World artiodactyls? Likewise, why did not early Oligocene horses make their way from North America into Eurasia, accompanied by the ubiquitous (to North America) oreodonts and the primitive camels? Very possibly the trans-Bering crossing, being a route through high latitudes, was not suitable for the intercontinental migratory movements of many mammals, especially herbivorous forms which may have been restricted by the distributions of the plants upon which they fed. The problems here are difficult and subtle, as we know from the study of modern mammals, and when one considers that it is necessary to view many extinct animals, especially the herbivores, without much knowledge of their basic diets, the difficulties are apparent. For example, we know from observation that today in Africa the local distributions of antelopes are frequently determined by the species of grasses available for grazing, or by the presence or absence of the tsetse fly. How might one ever make inferences about many details of mammalian distributions in past geological ages, assuming that factors similar to these might be involved? Such considerations should be kept in mind, even though they can be but of academic interest to us. We can only guess at why some animals, which to our eyes should have moved back and forth between the continental regions of early Oligocene Holarctica, nevertheless failed to extend their boundaries. The imponderables are there; we can only view the results. And the results were that early Oligocene time was marked by considerable admixtures, caused by mammals passing from Eurasia into North America, and likewise from North America into Eurasia, to mingle with those mammals that maintained comparatively stable distributions. So it was that the early Oligocene mammalian faunas of the north circumpolar land masses were related, by virtue of the migrants, and yet in many aspects distinctive, by virtue of the nonmigrants.

With the beginning of middle Oligocene history and from then until the end of the epoch it would appear, as has been mentioned, that the trans-Bering bridge was broken, with virtually no interchange of mammals between North America and Eurasia. Consequently the mammalian assemblages of these regions became increasingly distinctive, owing to their isolation from each other.

The Brulé beds of the White River sequence in North America, frequently divided into Lower Brulé, or Orella, and Upper Brulé, or Whitney, these subdivisions being indicative of middle and late Oligocene time respectively, are particularly marked by the abundance of oreodonts contained within them and the complete lack of titanotheres—giant rhinoceroslike herbivores so characteristic of early Oligocene time. For this reason the Middle and Upper Oligocene beds of the White River sequence have often been referred to as the "*Oreodon* beds," to distinguish them from the Lower Oligocene "*Titanotherium* beds." The fossil collector who traverses these upper levels in the White River Badlands will find a varied array of carnivorous mammals, including the massive and rather clumsy creodont, *Hyaenodon,* lightly built canids, weasellike mustelids, and early saber-toothed cats, exemplified by the well-known genus *Hoplophoneus,* perhaps comparable to a modern lynx in size, but armored with the enlarged saberlike canine teeth so characteristic of the group of cats to which it belonged. Also present in these sediments are small three-toed horses no larger than sheep, such as *Mesohippus,* lightly built, cursorial rhinoceroses, in general build more like horses to our eyes than to the rhinoceroses with which we are familiar, and large, heavy aquatic rhinoceroses, tiny deerlike herbivores, primitive peccaries, large entelodonts—which were piglike mammals comparable to oxen in size—the many-horned *Protoceras,* an artiodactyl like an antelope in body form, but having a skull provided with three longitudinally arranged pairs of horns, and, of course, the ubiquitous oreodonts. There were numerous small mammals—rodents, rabbits, and insectivores.

Faunas containing these mammals are rather different in character from the late Oligocene faunas of Asia, as found in the Hsanda Gol and Houldjin beds of Mongolia. There are, of course, certain resemblances, and in Asia one may find some mammals similar to those of North America—descended from ancestors that were common to the two regions during the earlier portions of Oligocene history. Such are some of the carnivores, including the long-lived and widely distributed creodont *Hyaenodon,* certain rodents, rabbits and insectivores, rhinoceroses, and piglike entelodonts. But in Asia there were no horses at this stage of earth history and, of course, none of those Oligocene mammals that, like the oreodonts and bizarre-horned protoceratids, were so peculiar to North America. In the Upper Oligocene beds of Asia are found gigantic, hornless rhinoceroses, as large as elephants in bulk, known as baluchitheres. And, strangely enough, the gigantic titanotheres still persisted in Asia, as contrasted with their absence, owing to extinction, in North America. These differences between the late Oligocene mammals of Asia and North America would seem sufficient to indicate that the land-living vertebrates of the two regions were developing independently of

FIGURE 85. A comprehensive view of some of the mammals comprising the rich and varied Upper Oligocene fauna of North America. These restorations are based upon fossils from the White River Badlands of South Dakota. 1. *Archaeotherium,* a giant piglike entelodont. 2. *Poebrotherium,* a small camel. 3. *Merycoidodon,* an oreodont. 4. *Agriochoerus,* an agriochoere. 5. *Mesohippus,* a three-toed horse. 6. *Hoplophoneus,* a saber-toothed cat. 7. *Bothriodon,* an anthracothere. 8. *Hyaenodon,* a carnivorous creodont. 9. *Hyracodon,* a small, running rhinoceros. 10. *Protoceras,* a horned herbivore. 11. *Caenopus,* a rhinoceros. The drawings are all to scale; the picture of the pointer dog gives a basis for comparisons.

each other, evidently with no trans-Bering bridge to allow migrations back and forth.

And when we progress on to the west, from Asia into Europe, there are differences that distinguish the European Middle and Upper Oligocene mammals from those of North America as surely as do those differences that have been outlined for Asia. Some are negative—the absence in Europe of horses and of those various Upper Oligocene mammals so characteristic of North America. Others are positive—the presence in Europe of mammals unlike those of North America, mammals of the sort that already have been mentioned in the discussion of early Oligocene Holarctica. Again, as in the case of Asia, it seems evident that Europe was out of contact with North America during this portion of earth history, and for the same reason—the interruption of the trans-Bering bridge.

OLIGOCENE MAMMALS OF NORTH AFRICA

It will be recalled from the foregoing chapter that Upper Eocene and Lower Oligocene mammals have been found rather abundantly in northern Africa, represented in the Fayûm district of Egypt. Whereas the Upper Eocene levels of the Fayûm beds, the Qasr-el-Sagha Formation, are only sparsely fossiliferous—containing the remains of moeritheres, the basic proboscidean ancestors, and a few scattered remains of other animals—the Lower Oligocene sediments in this sequence, the Jebel-el-Quatrani Formation, have yielded a considerable range of mammals peculiar to this one region. So it is that in the Lower Oligocene portions of the Fayûm beds the first of the mastodonts are seen, represented by the genera *Palaeomastodon* and *Phiomia.* These heavy-limbed, bulky mammals had four large tusks, two in the skull and two in the jaws, and obviously possessed trunks, in these and many other characters thus setting the basic patterns for proboscidean evolution, which was to be so very successful during the remainder of Cenozoic time. Also in these beds are sirenians, or sea cows, and giant hyracoids, related to the modern conies or dassies of the Middle East and Africa, and in addition a gigantic rhinoceroslike mammal, previously mentioned, with enormous paired horns rising from the top of the skull. This isolated and unique mammal, *Arsinoitherium,* is found in the Fayûm beds and nowhere else, and like the first proboscideans, the sea cows and the hyracoids, represents an evolutionary line that probably had its origins in North Africa.

Even though northern Africa was the place of origin for various mammals, some of which were to spread over the world during middle and late Cenozoic times, it was also an area invaded by mammals from the north. These are seen in the Lower Oligocene Fayûm beds as creodonts, including that wide-ranging genus *Hyaenodon,* and anthracotheres, which were early artiodactyls, or even-toed hoofed mammals, with four toes, some of them perhaps eventually to be ancestral to the hippopotamuses. The creodonts and anthracotheres probably entered this northern edge of Africa from Europe, showing that there probably was a land connection between northern Africa and early Cenozoic Laurasia at the end of the Eocene and the beginning of the Oligocene epochs.

Of particular interest in the Fayûm beds are the remains of small and ancient primates—including the earliest of the anthropoids. Among these are *Apidium*—a true Old World monkey—and *Propliopithecus*—an early ape. This latter genus gives us a glimpse of the remote ancestors of man.

The interesting mixture of autochthonous and immigrant mammals in the Lower Oligocene beds of northern Africa would seem to show that the evident connection between this continent and the lands to the north must have been one of limited nature—perhaps a restrictive filter bridge. Otherwise one might expect a larger representation of northern animals among the fossils found in the Jebel-el-Quatrani beds, and conversely, one might expect some of the proboscideans and hyracoids, so very characteristic of this stage of faunal development in Africa, to have appeared in the Oligocene of Eurasia. Such was not the case; our record of early proboscideans is restricted to northern Africa, and the same is true for the hyracoids. Not until the beginning of Miocene history did the proboscideans, in the form of primitive mastodonts, break away from the confines of their African homeland to populate other parts of the world. As for the hyracoids, they appear to have been an African group for a large part of their history.

There is as yet no knowledge concerning the development of Oligocene land-living vertebrates in other parts of Africa.

MIOCENE MAMMALS OF HOLARCTICA AND AFRICA

Two significant developments marked the opening of Miocene time in the northern hemisphere. One was the apparent reestablishment of the trans-Bering crossing, so long interrupted during Oligocene history, thus affording a bridge by which land-living vertebrates once again could cross back and forth between the western and eastern parts of the great circumpolar Holarctic region. The other was the continued modernization of mammalian faunas—a process that, of course, extended to those mammals inhabiting Africa as well as North America and Eurasia. This progressive development of Miocene faunas was the result in part of the expansion of mammalian families of modern relationships and in part of extinctions of the more archaic mammalian types. For example, the titanotheres, so prominent in early and middle Oligocene faunas, had disappeared, as had various other groups of mammals that formerly played important roles in the ecology of Oligocene land life. So it was that in North America, Eurasia, and Africa the Miocene faunas were composed largely of mammalian families still inhabiting the world. Only a fourth of the families comprising Miocene faunas in Holarctica and Africa are now extinct; consequently the faunas of the Miocene seem rather familiar to eyes accustomed to the mammals of the modern world.

The record of Miocene land-living vertebrates is abundantly preserved in North America, Eurasia, and Africa. In the high plains of western North America, for instance, a sequence can be followed from the Lower through the Middle to the Upper Miocene, involving in order the Arikaree and Harrison beds of Nebraska, and the John Day sediments of Oregon, all of early Miocene age; the Sheep Creek, Marsland, Snake Creek, and Hemingford beds of Nebraska of middle Miocene age; and the Pawnee Creek beds of Colorado and the Barstow Formation of California, of late Miocene affinities. These are only a few of the numerous Miocene mammal-bearing formations in North America; to list them all would lead us into diversions unnecessarily removed from the purpose of this discussion.

In Europe one can follow the Miocene from its earliest to its latest stages in the deposits of St. Gerand le Puy in France, Vallés Penedés in Spain, and others, representing the Lower Miocene; Sansan and Simorre in France, Steinheim and Oenigen in Germany, and Monte Bamboli in Italy, of middle Miocene age; and Grive St. Alban in France and Sebastopol in Russia, with Upper Miocene faunas.

In Asia the Hsanda Goł beds in part and the Loh Formation of Mongolia, as well as the Bugti beds of Pakistan are of early Miocene age; the Kamlial of the Siwalik Hills of India are of middle Miocene age; and the Mongolian Tung Gur Formation represents the Upper Miocene.

Finally, in Africa, Lower Miocene sediments are represented in northern Africa by the Moghara beds of Egypt, and in eastern Africa by the Rusinga and Lake Rudolph beds of Kenya. In Kenya are the Fort Ternan beds of late Miocene age.

From such an array of sediments (and, of course, this is only a partial list) comes the evidence as to the vertebrate faunas, dominantly mammalian, that inhabited the northern hemisphere and Africa during Miocene times. These faunas are of interest because of the interrelationships between them—the result of land connections between the northern continents and Africa, which at that time were closely approaching their present positions.

One might suppose that with the joining once again of North America and Eurasia by the reestablished trans-Bering bridge, there would have been a rapid interflow of mammals between the two continental masses. But such was not the case. There was definitely an interchange of mammals, from west to east and in the opposite direction, but it was strangely limited during the earlier phases of Miocene history. Consequently the faunas of Eurasia and North America evolved on the whole rather independently during early and middle Miocene time, as they had evolved during middle and late Oligocene time. On the two sides of the Bering region the mammalian assemblages continued with many of the same elements that we have seen as characterizing them through much of Oligocene history.

In Eurasia there were civets, or viverrids, giant rhinoceroses, pigs, and anthracotheres, as there had been during the Oligocene epoch. Added to these indigenous elements now were early mastodonts—generally with four tusks, two in the skull and two in the lower jaw—as well as the strange dinotheres—proboscideans with large, recurved tusks in the lower jaws and no tusks at all in the skull. It was during this time that the giraffes and the antelopes, which were to enjoy such an exuberant evolutionary history in the Old World, made their appearance. Of particular significance was the presence of early anthropoids, such as *Dryopithecus,* in many respects very similar to a modern chimpanzee, probably descended from African ancestors. In North America there were large, short-limbed rhinoceroses, various rodents not found in Eurasia, and numerous descendants from indigenous Oligocene ancestors, such as peccaries, oreodonts, camels, and other artiodactyls, notably the antilocaprids—a group that throughout its history was confined to North America and is represented today by a single survivor, the pronghorn.

Yet in spite of the evident differences between the mammals of

North America and Eurasia, as a result of the separate developments of faunas in the two regions, there were certain interchanges that indicate an open way for the intercontinental movements of animals between the two hemispheres. Still it is indeed puzzling that more mammals did not make the crossing back and forth during early and middle Miocene history.

For example, by middle Miocene time mastodonts had appeared in North America, quite obviously reaching this continent from a Eurasian base. Such animals could only have come across from the Old World into the New World by way of a bridge that almost certainly was the trans-Bering bridge. The same can be said for early deer (as opposed to the hypertragulids and other indigenous "deerlike" North American artiodactyls), which surely were immigrants from Eurasia. In the other direction, early beavers crossed from North America into Eurasia. Also, and of particular importance, was the migration from North America into the Old World of *Anchitherium,* a horse of primitive structure, descended from and retaining many of the features in enlarged aspect of *Mesohippus,* the characteristic North American Oligocene horse. Strangely enough, other horses that were evolving in North America at this time did not make the crossing into Asia and Europe.

And so we see the anomaly of an incompletely utilized trans-Bering bridge. Why it was so widely ignored by many mammals at each end of the crossing is, as has been said, a puzzle. But so it was, until late Miocene time. And then the interchange of mammals between North America and Eurasia increased to some degree.

In these final years of the Miocene various carnivores crossed back and forth between the two continents, among them the "bear-dogs," heavy, gigantic canids as large as bears, and certain cats, from Eurasia to North America, and the "hyenalike" dog *Borophagus,* a true canid, but one with a very heavy skull and jaws and massive teeth, in the other direction. But the herbivorous mammals on the whole still remained in their ancestral domiciles. What is especially interesting is the fact that proboscideans in the form of mastodonts of various types continued to flood into North America from Eurasia, while horses of conservative aspect, represented by the genus *Hypohippus,* a fairly large horse that retained the low-crowned teeth and the three-toed feet of its ancestors, crossed into the Old World. Yet the more progressive horses, which were evolving in great variety and numbers in North America, still failed to make the crossing. In fact, one of the strange characteristics of the mammalian interchanges between the Old and New World during late Miocene time was the scarcity of horse migrants and the abundance of adventurous mastodonts. Perhaps the high latitude position of the trans-Bering bridge was such that it did not favor the passage of grazing animals, which feed upon grass, but was suitable to a degree for browsers, which feed upon leaves of trees and bushes, such as the mastodonts must have been.

The movements of mammals between Eurasia and North America across the trans-Bering bridge continued from late Miocene times into the beginning of the Pliocene epoch, when the horse *Hipparion* invaded the Old World, where it characterized mammalian faunas across many thousands of miles of terrain. This is a development shortly to be discussed; at the moment it is mentioned to point out the fact that intercontinental movements at this stage of Cenozoic history probably took place in a series of waves or impulses, following one another in rather close succession, thus causing the development of faunas on the continents to follow in close sequence—often to the confusion of paleontologists who are trying to decipher them.

While these interchanges were taking place in the northern hemisphere, together with the often surprising lack of interchanges, there were also movements of animals between Africa and Eurasia. Of course, the great Miocene spread of mastodonts and dinotheres across Eurasia, with the extension of the mastodonts into North America, had its origins in Africa. As has been mentioned, these truly wonderful mammals seemingly were confined to Africa until the advent of the Miocene epoch, when they quickly invaded the lands to the north, to evolve in great variety. One of the distinguishing features of Miocene (as well as Pliocene) faunas in Eurasia and North America is the dominating position that the mastodont proboscideans occupy in these faunas. And in the Old World the ubiquitous mastodonts are supplemented by the unique dinotheres—which evolved along a singularly restricted line (confined to a single genus) yet nonetheless a widely distributed and persistent one, covering Eurasia and Africa, and ranging in time from the beginning of the Miocene until well into Pleistocene years.

The proboscideans were not the only emigrants from Africa into other parts of the world; advanced primates, especially anthropoids, also would seem to have arisen in Africa, to cross into Eurasia. *Dryopithecus* (already mentioned) is particularly noteworthy—a chimpanzee-like ape that became widely established in Eurasia during Miocene time, yet never made the crossing into the New World.

FIGURE 86. A composite scene showing some North American Miocene and Pliocene mammals. On the left is the Miocene chalicothere *Moropus*, a mammal related to horses and rhinoceroses, but with clawed toes, rather than hooves. Next to it is the small Miocene rhinoceros *Diceratherium*. The other mammals, of Pliocene age, include the three-toed horse *Hypohippus* in the foreground, the large "bear-dog" *Hemicyon*, the tusked mastodont *Trilophodon*, and the antelopelike hypertraguloid *Synthetoceras*.

As contrasted with these animals (all of them, interestingly, mammals of unusual intelligence), the hyracoids remained in Africa. Today hyraxes are found in the Middle East, but this probably represents a fairly recent incursion from the African continent into a nearby land.

In addition to the indigenous elements—proboscideans, hyracoids, and advanced primates—typical of African Miocene faunas, the mammalian assemblages of this continent included many animals found also in Eurasia. Such were the early rhinoceroses, anthracotheres, pigs, and primitive antelopes. All in all, it is evident that there was a considerable amount of exchange between Africa and Eurasia during the course of Miocene history.

By the end of the Miocene epoch the mammalian faunas of the northern hemisphere and of Africa were well established along patterns that were to persist and to distinguish them during the geological years to come. There were many aspects of these faunas unique to the continents they inhabited: early apes, civets, dinotheres, giant rhinoceroses, pigs, giraffes, and primitive antelopes in Eurasia, and many of these in Africa as well; hyracoids restricted to Africa; oreodonts, camels, deerlike hypertragulids, and antelopelike antilocaprids in North America. But along with such restricted animals there were widely distributed mammals, such as mastodonts, varied carnivores, certain horses, and rhinoceroses, giving evidence of wide movements between the continents. These movements, becoming increasingly abundant during the course of Miocene history, continued into the Pliocene, as we shall see.

HIPPARION AND THE ADVENT OF PLIOCENE TIME

The history of the earth and of life on the earth has been a continuing development, extending through billions, millions, and thousands of years of time. Our subdivisions of the long flow of physical and biological evolution are arbitrary, based to a large degree upon gaps in the record of time and life—gaps that would seem to form natural breaks at which lines of demarcation can be drawn. The artificiality of such lines is nicely illustrated by the attempt to distinguish a border between Miocene and Pliocene history, especially as it is recorded in the development of the mammalian faunas of the Holarctic and African regions. The fossil record is abundant, demonstrating the continual development of faunas, and the problem as to where to draw a line between the faunas typical of the end of the Miocene and those characteristic of the beginning of the Pliocene is to a considerable degree one to be solved by the personal bias of the authorities who have devoted their attention to the vertebrate life of late Tertiary age. There are difficulties, and any proposal put forward is open to objections.

It has been widely accepted, however, that perhaps the beginning of the Pliocene (at least in the Old World) may be signaled by the appearance about 12 million years ago of the horse *Hipparion* in mammalian faunas. This is at least a useful criterion.

Hipparion originated in North America and was descended from late Miocene ancestors. This horse, about the size of a rather large pony, was progressive by reason of its very tall cheek teeth for the grinding of harsh grasses, which is an advanced equine character, and its long legs for rapid running over hard ground, again an advanced character. But *Hipparion* was conservative in that it retained the lateral toes on the feet, so that each foot was composed of a large central hoof, with small accessory hooves, one on each side of the central toe. Horses of this type appear in North American sediments considered to be of basal Pliocene age.

It would seem that immediately after its appearance in North America *Hipparion* crossed the trans-Bering bridge to populate Eurasia. Its spread must have been wide and rapid, for almost simultaneously the so-called *Hipparion* faunas become evident throughout the extent of the Eurasian continent.

(Parenthetically, we have a modern example of the spread of horses to give us some insight into the conquest of Eurasia by *Hipparion*. When the Spanish conquistadors entered the New World, at the beginning of the sixteenth century, they brought Spanish horses into a horseless continent—the native American horses having been extinct for several thousand years. Some of the Spanish horses escaped, and probably some were also stolen by Indians. Within a matter of a few decades wild horses, the descendants of their Spanish equid forebears, populated the plains of western North America. By the late seventeenth and the eighteenth century the western Indians, formerly relatively sedentary people, had become nomadic horsemen, who thought that the horse had always been with them.)

Hipparion is present in the Clarendon beds of Texas, thus signaling the advent of Pliocene mammalian life in this part of North America. Comparable sediments are the Burge and Valentine beds of Nebraska, the upper part of the Santa Fe sequence in New Mexico, and the Alachua beds of Florida.

The *Hipparion* faunas of the Old World are found in the Concud beds of Spain, the Mount Leberon sediments of France, the Eppelsheim beds of Germany, and particularly the Pikermi and Samos beds of Greece, in which are found remarkably abundant faunas of early Pliocene age. In India are the Chinji and the overlying Nagri Formation, in Persia the Maragha beds, and in China the Lower Pliocene sediments of Honan and other regions. *Hipparion* faunas appear as well in North Africa, along the southern border of the Mediterranean.

From these widely distributed occurrences we see the tangible fossil record showing how *Hipparion* invaded the Old World, to become established through the length and breadth of Eurasia and into the northern border of Africa. This horse may have pushed farther into the African continent, but as yet we have no record of a more southern extension of its African range. To counter the push of *Hipparion* from North America into Eurasia, there was an opposite movement of mastodonts, especially the so-called shovel-tusked mastodonts (in which the lower jaw was very broad, to accommodate wide, flat lower tusks, the whole structure being remarkably like a large scoop shovel), from Eurasia into North America. Here is a stage in the history of Holarctic mammalian faunas marked by the back and forth movements of horses and mastodonts.

Strangely enough, however, there was little intercontinental migra-

FIGURE 87. At the beginning of the Pliocene epoch the three-toed horse *Hipparion* spread from North America, where it had originated from late Miocene ancestors, and crossed a trans-Bering bridge to populate the Old World.

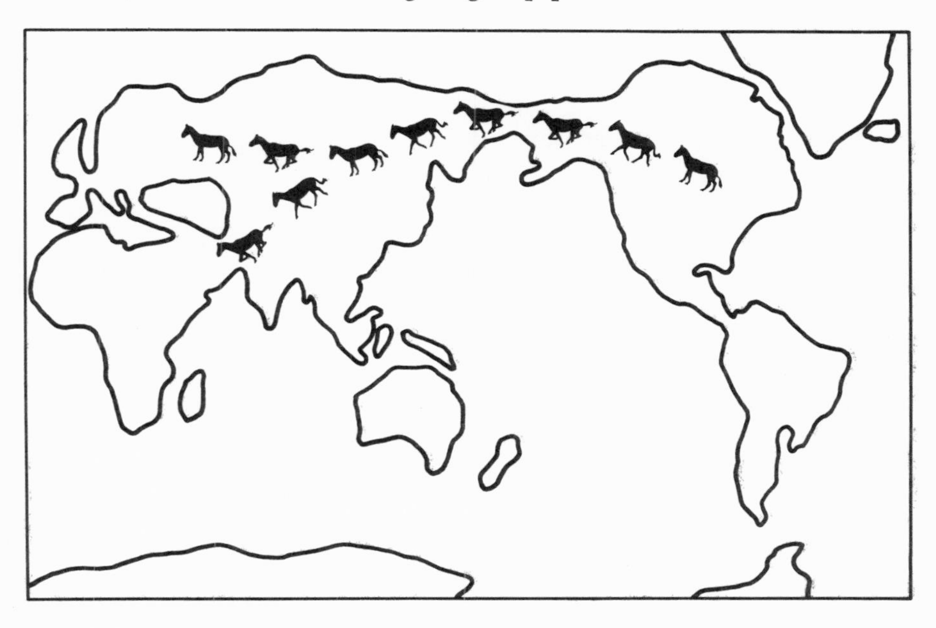

tion of other mammals during these initial years of Pliocene history. It would seem that the early Pliocene incursion of *Hipparion* across the trans-Bering bridge into the Old World and of mastodonts into the New World was a weak migratory pulsation following the intercontinental migrations of late Miocene time. Or perhaps it represents a tapering off of the Miocene migration. However that may be, the other mammals comprising the Lower Pliocene faunas of North America and Eurasia were very much confined to their respective regions, even though they were evolving and even though many of them would seem to have been perfectly capable of making the northern crossing. Of especial significance was the remarkable evolutionary burst of antelopes and early cattle in Eurasia. These, the most advanced and well adapted of the grazing mammals, embarked upon numerous, widely divergent adaptive lines that were to mark their success from then until modern times. Yet their evolutionary development was strictly an Old World phenomenon.

LATER PLIOCENE MAMMALS

The *Hipparion* faunas of the New and Old World, so successful during the span of early Pliocene history, gradually gave way to the equally successful and varied mammalian faunas of middle and late Pliocene times. And *Hipparion* frequently continued into the later phases of Pliocene history. Other horses appeared in North America, so that during the later years of the Pliocene the North American continent was the home of many parallel lines of horse evolution. Here, as well as in the Old World, the mammalian faunas were numerous and rich.

Typical of the middle and late Pliocene mammal-bearing sediments of North America are the Hemphill beds of Texas, the Ash Hollow of Nebraska, the Rattlesnake formation of Oregon, and the Thousand Creek beds of Nevada. In Europe there are the sediments of Montpellier and Perpignan, France. The Dhok Pathan Formation of India would seem to be a long-range sequence of sediments, embracing perhaps a part of the Lower Pliocene, with *Hipparion* present, and virtually all of the Middle and Upper Pliocene as well. Sediments of middle or late Pliocene age are found in northern Africa and in the southern part of the continent also, along the Vaal River.

From the abundant mammals found in many of these deposits it is evident that the intercontinental movements across the trans-Bering bridge, so restricted during early Pliocene history, became much intensified in middle and late Pliocene times. There was, it would seem, a new spurt in the exchange of mammals between North America and Eurasia. This faunal interchange, however, would seem to have largely involved the rodents and the carnivores; the herbivorous mammals still remained for the most part in their ancestral homelands.

Thus middle and late Pliocene years saw a great influx of carnivores from Eurasia into North America. Bears crossed the bridge into the New World, as did badgers, skunks, and saber-toothed cats of Old World origin. This migration included also martens, otters, various other cats, and even some hyaenids—although these last never became abundantly established in the New World. Along with this small army of carnivores there came into North America during the later years of the Pliocene certain short-jawed mastodonts, differing from their progenitors in that the lower jaw had become shortened and had lost its tusks, and deer. In the other direction various rodents, including advanced beavers, crossed into Eurasia.

Yet in spite of this increased rate of mammalian exchange between North America and Eurasia, the bulk of the faunas in the two continents evolved *in situ,* as they had in previous years. Eurasia and Africa continued to be the home of monkeys and apes, of Old World porcupines, civets, clawed chalicotheres, and rhinoceroses (these perissodactyls having migrated between the two continents in earlier times), pigs, anthracotheres, tragulids (the small primitive deerlike artiodactyls now so characteristic of Africa and the Orient and from which in earlier Tertiary times the first deer had their origins), giraffes, and the burgeoning antelopes and cattle. North America continued to be a great arena for the evolution of horses, and of peccaries, camels, and antilocaprids (pronghorns) as well. The Pliocene raccoons were also North American residents, although it would seem probable that some of them had migrated into Asia at an earlier date, to give rise to the pandas of that region. And in North America there were some new mammals—edentates from South America—which never reached the Old World. A few rhinoceroses lingered on into late Pliocene times in North America, the remnants of a long heritage on this continent, but the chalicotheres, so well developed in some of the Miocene faunas of the North American continent, were now extinct in this region.

Even though the faunas of North America and Eurasia were distinct by reason of these numerous mammals confined to the two continents, they were not so distinct as they had been in earlier times. The increased intercontinental movements of various mammals lent an increased degree of similarity to the faunal assemblages on the two sides of the trans-Bering bridge. Moreover, these resemblances, resulting from intercontinental migrations, were destined to expand. For the movements of mammals across the trans-Bering bridge continued—from this time until the middle of the Pleistocene epoch, the time of the great northern glaciers. Indeed, it would appear that there was a continuous interchange of mammals between North America and Eurasia throughout this particular span of geologic history, and this interchange, as we shall see, involved many genera of mammals still surviving, belonging to families of mammals that for long had been common to the two regions. It was an interchange that would seem to have reached its culmination during the transition from the Pliocene into the Pleistocene epoch, the stage of faunal evolution marked by the waxing of the Blancan faunas of North America, the Villafranchian faunas of Europe, and the Nihowan fauna of China, to which mammalian assemblages we will soon turn our attention.

EMERGENCE OF THE ISTHMUS OF PANAMA

In the meantime, however, it is necessary to note a geological development of late Pliocene date that affected the composition of mammalian faunas in North America. This was the reestablishment of the Panamanian isthmus, once again securing the connection between North and South America, that according to one theory had been broken in early Cenozoic time. This naturally inaugurated a pathway in the form of a filter bridge, whereby some of the land-living animals in the two continents could make their ways back and forth, north and south. As for the influx of North American mammals into South America, we need only to note here that it occurred with considerable strength; it will be described in the next chapter. Our concern at the moment is with the movement of South American mammals into North America during the final years of the Pliocene. This south-to-north invasion was a movement of limited scope. It would appear that many of the South American mammals, although seemingly capable of making their way north, were unable to do so because in a structural, evolutionary sense they were not sufficiently advanced to displace their North American counterparts, already well entrenched in their homeland. However, members of one group of South American mammals were successful in making their way north, and for the remainder of Pliocene time and through the Pleistocene until what may be termed "subrecent time," they inhabited much of North America. These were the giant edentates —the huge armored glyptodonts and the massive ground sloths. The small cousins of the glyptodonts, the armadillos, also came into North America from the south, to survive into modern times.

These migrations along the Panamanian isthmus added new elements to the North American faunas of late Pliocene times, but they failed to extend the dominions of the giant edentates beyond North America. They did introduce exotic animals quite different from those that had lived in North America during the major span of Cenozoic time. The late Pliocene edentates were the culmination of a long development in the isolated continent of South America, of evolution in isolation that was shared by various other groups of mammals. To this part of the world, and to the other Cenozoic island continents of the southern hemisphere—Australia and Antarctica, and perhaps peninsular India—we will now turn our attention before continuing with the story of Ice Age continents and mammals, the story of a world of land and life immediately antecedent to our world of today.

X. ISLAND CONTINENTS AND ISOLATED FAUNAS

THE ISOLATION OF SOUTH AMERICA

At the beginning of Cenozoic time South America was populated by various primitive mammals (as described in Chapter VIII), many of them, according to one view, immigrating into that continent from North America along an ancient Panamanian isthmus, or according to another probably originating on that southern land mass from very primitive ancestors. The primitive ancestors included ancient marsupials, which had appeared during Mesozoic time and had become rather widely spread throughout the earth. The possible immigrants from North America would seem to have included some primitive hoofed mammals which were to give rise to the host of strange hoofed mammals that evolved on the South American continent during its isolation from the rest of the world, as well as some early ancestors of the edentates—the sloths, glyptodonts, armadillos, and anteaters, which were to become so prominent in South American faunas. As already mentioned, it has been suggested that even these mammals arose from Cretaceous ancestors in South America. So it was that during Paleocene and Eocene times a base was established for South American mammalian faunas, to be composed of marsupials, edentates, and a varied array of hoofed mammals quite different from the hoofed mammals that inhabited other continents.

Then, during Eocene time, if not before, South America became essentially an island continent. It probably was not completely isolated from other land masses, for modern evidence indicates that there probably was a connection between South America and Antarctica by way of the Scotia Arc, and that in turn Antarctica was connected with Australia until the Eocene epoch. But certainly South America was isolated from those northern lands and from Africa, where mammalian faunas were evolving in profusion and diversity.

Consequently the South American mammalian faunas developed during much of Cenozoic history along lines quite different from but frequently parallel to those that were evolving in other parts of the world. Indeed, the evolution of Tertiary mammals in South America provides a nice example of how animals of differing origins respond through evolutionary processes to the habitats in which they live. Such responses may produce results quite different from anything to be seen elsewhere; they may result in parallelisms that resemble unrelated animals in other regions to a truly astonishing degree.

These South American mammalian faunas are preserved within a sequence of sediments found for the most part in Patagonia, in southern Argentina, ranging from the Paleocene into the Pleistocene, thereby affording a rather complete record of mammalian life on this southern continent. At the base of the sequence is the Rio Chico Formation of Paleocene age, this followed by the Casa Mayor and Musters formations of Eocene age. Above the Musters beds are the Oligocene Deseado and Colhué Huapí formations, and following them the Miocene Santa Cruz Formation. This latter formation has yielded a varied and abundant fauna, giving us an excellent view of mammalian life in South America during early Miocene time. The later Miocene is represented by the Chasico and Rio Frias sediments of Argentina and by the La Ventana Formation of Colombia, from which excellent fossils have been collected in recent years. The Lower Pliocene is represented by the Entre Rios and Catamarca beds of Argentina, and the Upper Pliocene by the Monte Hermoso Formation, also in Argentina. Finally there are the Argentinian Chapadmalal and Pampas beds, these being fossil repositories from which large quantities of Pleistocene bones have been excavated, as well as the Pleistocene Tarija sediments of Bolivia. All in all the South American sequence provides an interesting and instructive record of Cenozoic mammalian life, largely unique to this one continent.

Incidentally it is pertinent to note here that young Charles Darwin excavated huge toxodont and edentate bones from a riverbank in Patagonia during the famous voyage of the *Beagle*. And these fossils were crucial to the development of his concept of evolution.

INDEPENDENT EVOLUTION OF SOUTH AMERICAN MAMMALS

Many of the marsupials that evolved during Tertiary time in South America retained the primitive nature of their Cretaceous ancestors; in short, they evolved as opossums and similar animals—these being essentially Cretaceous marsupials that have continued through long geologic ages into modern times. But one group of South American marsupials, the borhyaenids, developed as highly adapted carnivores, making their living by preying upon the various animals, large and small, with which they were contemporaneous. The first borhyaenids appear in Rio Chico sediments as comparatively primitive or generalized carnivorous marsupials. By Santa Cruz time there were large borhyaenids, rather wolflike in form and presumably in habits, and in the Pliocene is an amazing genus, *Thylacosmilus,* a catlike marsupial as large as a leopard in which the skull is armed with huge saberlike canines, even more extreme in their specialization than those of the well-known saber-toothed cats of the northern hemisphere.

It has been suggested that the borhyaenids of South America had an ancestry in common with that of the carnivorous marsupials of Australia, represented today by the so-called Tasmanian wolf, or thylacine, *Thylacinus,* of Tasmania. Certainly there is a strong resemblance between the thylacine, which shows an uncanny resemblance to a wolf or dog in spite of its marsupial heritage, and some genera of borhyaenids, which lends some credence to the possibility that these animals crossed between South America and Australia by way of the intermediate Antarctic continent. Perhaps future work will throw some light on this possibility. On the other hand, it is equally or even more probable that the resemblances between the borhyaenids and thylacines are the result of parallel evolution.

The edentates, starting from modest beginnings, evolved into the armadillos, sloths, anteaters, and the gigantic armored glyptodonts and ground sloths—more or less like armadillos and sloths on a grand scale—all animals almost without parallel in other parts of the world. And for this reason it is difficult to equate their ecological roles with those of mammals on other continents. These animals were prominent in the

Cenozoic faunas of South America, as indeed the surviving edentates are today.

Of particular interest are the hoofed mammals of Cenozoic South America, represented by the litopterns, astrapotheres, pyrotheres, and notoungulates, all long since extinct and all quite unfamiliar to anyone who has not been initiated to the study of South American fossil mammals. The litopterns evolved along several radiating lines, among which some of their more advanced members came to resemble in a vague way the camels of the northern hemisphere. The late Cenozoic genus *Macrauchenia*, for example, is rather like a large camel and in life was provided with an elongated, flexible nose, or proboscis. A particularly striking case of parallelism among the litopterns is to be seen among certain genera, such as *Diadiaphorus* and *Thoatherium*, of Miocene and Pliocene age, which evolved along a line that was uncannily similar to that of the Tertiary horses in North America in the development of the head, body, and limbs. Indeed, some of the litopterns of this ilk outhorsed the horses in the perfection of a single-toed foot, for rapid running over hard ground.

The largest and most varied group among the South American hoofed mammals comprised the notoungulates, ranging in age from the Paleocene into the Pleistocene, and in size and adaptations from small rabbitlike mammals, known as "typotheres," to huge, heavy-bodied grazing mammals, known as "toxodonts." It would require far too much space in a book such as this to try to describe the multitudinous notoungulates; suffice it to say that between them and the litopterns many of the ecological niches occupied in other parts of the world by the perissodactyls and artiodactyls were successfully filled.

The astrapotheres were large, rather clumsy ungulates, in which the skull was frequently provided with a pair of saber teeth. The pyrotheres were gigantic forms (known at this date largely from skulls and jaws), which imitated in a way the proboscideans of Holarctica and Africa. In fact, as has been mentioned, many of the earlier students of South American fossil mammals thought that the pyrotheres were proboscideans; they have teeth like those of the dinotheres, and it is obvious from the structure of the skull that in life they had trunks of considerable size.

SPORADIC INCURSIONS FROM THE NORTH

These mammals of South America, evolving in complete isolation from the rest of the world, were augmented by some other mammals

FIGURE 88. South American mammals at the end of their long isolation, when autochtonous forms still predominated, but when there already was an admixture of immigrants from the north. These mammals are of Pampean (Pleistocene) age. 1 and 2 are glyptodonts (*Daedicurus* and *Glyptodon,* respectively), gigantic armored edentates, related to the modern armadillos. 3. *Macrauchenia,* a litoptern that fulfilled roughly the ecological role of a camel, before South America was invaded by the little camels known today as llamas and their relatives. 4. *Hippidion,* a horse descended from earlier horses coming in from North America. 5. *Toxodon,* a notoungulate—one of the last of its line. 6 and 7 are giant ground sloths, related to the modern tree sloths. (*Megatherium* and *Mylodon,* respectively). The pointer dog gives the scale.

that rather accidentally reached South America from distant continents. Rodents appear in South American sediments of early Oligocene age, and from this base these most numerous of mammals evolved in South America during the remainder of Cenozoic time, to populate the continent with abundant rodents different from those in other continents. Again we see evolution in isolation, but from beginnings that clearly may be related to animals elsewhere. Similarly, monkeys are found in late Oligocene sediments, and from them there evolved the typical South American monkeys, differing in various details from the monkeys of the Old World. Again, procyonids, related to the raccoons of North America, suddenly appear in Miocene deposits in South America. The sudden and rather random appearance of these few mammalian groups in South America would seem to indicate that they reached the South American continent adventitiously as waifs, the original immigrants very likely floating to coastal landings on masses of vegetation or on logs.

The relationships of the early rodents and primates of South America indicate that they were very probably of North American origin, being descended from Eocene ancestors that lived on the northern continent. It is not beyond reason to think that these animals lived in Middle America, as it then existed during late Eocene time, and that some of them made their way along an island chain from their original homeland to the South American coast. The procyonids probably reached the continent in the same way, at a later date.

THE PANAMANIAN FILTER BRIDGE AND USURPERS FROM THE NORTH

Such were the South American mammalian faunas as the Cenozoic drew toward its close—faunas composed of marsupials, edentates, the several groups of hoofed mammals unique to the continent, and rodents, monkeys, and some raccoonlike procyonids. Then with the reestablishment of the Panamanian isthmus, during the Pliocene epoch, there was a massive invasion of South America along this filter bridge by mammals from the north.

It was in effect a long parade, beginning in the Pliocene and continuing into the Pleistocene epoch. Some insectivores entered South America, along with rabbits, squirrels, and mice. Also among the invaders were wild dogs of various kinds (using the word "dogs" in a very broad sense), bears, more raccoons, weasels, and large and small cats. In addition, mastodonts, so numerous in North America, entered the southern continent. Also making the crossing along the isthmian bridge were horses, tapirs, peccaries, llamas, and deer.

We have seen that there was a northward movement along the filter bridge at the same time, with the result that ground sloths and glyptodonts, armadillos, capybaras, and porcupines entered North America. The movement of mammals from south to north was limited as contrasted to the movement in the other direction, and its effects were to augment the faunas of North America. But the movement of mammals from north to south was overwhelming and, from the point of view of the native South American mammals, catastrophic. The old inhabitants of the southern continent were crowded and eventually became extinct before the onslaught of the northern invaders. Some of the long-established inhabitants held on into Pleistocene time, but in the long run it would seem that, in spite of their remarkably successful adaptations to South American habitats, the results of evolution through many millions of years, they were unable to withstand the pressures from the northern mammals. In brief, the mammals from North America would appear to have been more "efficient"—however such a term may be interpreted—than the animals they displaced.

The marsupial borhyaenids quickly succumbed to competition from the obviously more intelligent placental carnivores and became extinct during late Pliocene time. The litopterns and notoungulates continued into the Pleistocene epoch, and then they dwindled away, almost surely as a result of competition from such efficient browsers and grazers as horses, tapirs, peccaries, and deer. The ground sloths and glyptodonts persisted into late Pleistocene time before becoming extinct, as they did also in North America. Just what may have been the factors leading to the extinction of these edentate giants is a question difficult to answer.

So within a relatively brief span of geologic time the invaders from the north usurped the ecological niches long held by the varied large mammals of South America, to share the continent with those earlier inhabitants that did not give way before the wave of immigrants—the smaller marsupials (such as various opossums), the rodents, and the monkeys. Thus the South American or Neotropical mammalian fauna as we know it became established, not as a result of distributions inherited from drifting continents, but rather as a result largely of invasions along a filter bridge. It is a story distinctly in contrast to the history of mammals on the other continents.

AUSTRALIA DETACHED, AND MARSUPIAL EVOLUTION

A story different from but parallel to that of South American mammalian faunas is revealed by the recent and fossil mammalian faunas of Australia. Today the native fauna of Australia and adjacent islands is composed of marsupials, of some monotremes, and of such placental mammals as rodents, bats, and the dingo. It seems evident that the rodents came into Australia adventitiously, as they did to South America. There is no mystery as to the presence of bats in Australia. It is obvious that the dingo was brought to the continent by aboriginal immigrants. The bulk of the fauna, however, composed of a great host of marsupials of all kinds and of two monotremes—the duckbill platypus and the spiny echidna—represents a long history of evolution in isolation upon the Australian continent, as has been partially revealed in recent years by fossil evidence.

It is clear that Australia, which probably separated from Antarctica during Eocene time, to drift through the intervening years to its present position, was after its separation truly an island continent. Before it had become separated it was populated by monotreme and marsupial mammals, but for some reason the placentals never obtained a foothold there. Consequently the island continent, as it drifted from its original southerly position contiguous to Antarctica (as this latter continent was then located) to its present less southerly position, was a haven for marsupials—free from any placental competition. This was fortunate for the marsupials, because we can see that in those lands where marsupials and placentals live together, the marsupials (even though frequently successful) are still subordinated by the much more intelligent and perhaps more adaptable placental mammals. The unencumbered marsupials of Australia therefore evolved along numerous lines, to occupy the ecological niches that on other continents are generally held by placental mammals.

Consequently the modern fauna of Australia is made up of marsupials that perform the function of mice and rats (although in this respect they must compete with the placental rodents that had invaded Australia—probably during the Pliocene epoch), of squirrels, of predatory carnivores, and of varied herbivores. These marsupials are represented by such forms as the many opossums, dasyures, phalangers, bandicoots, the koala, the wombat, the Tasmanian devil, the thylacine or Tasmanian "wolf," and an almost overwhelming and confusing host of wallabies and kangaroos. Many of the opossums and phalangers may be equated with mice and squirrels on other continents, the dasyures with cats, the bandicoots with insectivores and rodents, the thylacine with a wolf or fox, the Tasmanian devil perhaps with a wolverine, the koala in a vague way with tree sloths, and the wallabies and kangaroos with the browsing and grazing mammals, the perissodactyls and artiodactyls, in other parts of the world.

The egg-laying mammals of Australia—the platypus and the echidna—probably represent holdovers from Mesozoic times; indeed, they are in a sense persistent Mesozoic mammals, or perhaps even persistent therapsid reptiles, if you wish to look at them in this way. They occupy niches of their own; the platypus is an aquatic animal, living in rivers

FIGURE 89. Modern Australian marsupials, which having evolved in isolation on a drifting island continent, occupy ecological niches filled in other lands by placental mammals. The Tasmian "wolf," or thylacine (now virtually extinct), was an aggressive predator of rather doglike appearance. The wombat is more or less like a marmot, the phalanger or opossum like a squirrel, the Koala vaguely like a tree sloth, the bandicoot like a rabbit, and the "mouse" and "mole" like the placental mammals from which they take their vernacular names.

and feeding on aquatic worms, prawns, and molluscs, while the echidna is an insectivore, protected by spines, like a hedgehog or porcupine.

The fossil mammals found in Australia are instructive but not particularly revealing concerning the locus of their ancestry. They show us that marsupials have been evolving on the continent since at least Oligocene times, which is no surprise, but they do not indicate the ultimate origins of the Australian mammalian fauna. It does seem probable that marsupials have been in Australia since late Cretaceous time: why, then, are placentals absent, since placental mammals were also well established in other regions at this stage of geologic history? Perhaps there were connections between South America and Australia through Antarctica in late Cretaceous or early Tertiary times, allowing marsupials to pass back and forth along some sort of a filter bridge, which the placentals failed to utilize. It will be interesting to see if future explorations in the Cretaceous or early Cenozoic sediments in Antarctica throw any light on this aspect of the problem.

It is possible but perhaps less likely that the first marsupials reached Australia accidentally as waifs that had drifted in stages to the coast from Asia, along a series of intermediate islands. If this were the case, then the absence of placentals might be explained as a matter of pure chance; it just happened that Cretaceous marsupials drifted across a series of narrow oceanic barriers on logs or floating vegetation, and Cretaceous placentals did not.

As for egg-laying monotremes, it may be that these animals are the remnants of therapsid reptiles, or very early mammalian descendants of therapsids, that have been in Australia since Triassic or early Jurassic times.

PENINSULAR INDIA AND ANTARCTICA—WHAT WE DO NOT KNOW

Antarctica, once an integral part of Gondwanaland, was drifting toward its present isolated polar position during Cenozoic time. As has been stated, it probably was joined with Australia as late as the Eocene —but then the junction was ruptured, and the two continental masses followed their separate courses. As also has been stated, it is likely that there was some sort of a connection between South America and Antarctica during a considerable fraction of Cenozoic history. As yet there is no fossil evidence bearing upon the problem of Cenozoic land-living vertebrates in Antarctica. The recent discoveries of rather abundant Triassic amphibians and reptiles in the Transantarctic mountains may be a portent of the future; there very possibly are many other fossil faunas to be found on the south polar continent. Field studies in Cretaceous and Cenozoic sediments, particularly in the Antarctic peninsula, may reveal as yet unknown facts concerning the distribution and development of early mammalian faunas in this part of the world.

One other region to be considered at this place is peninsular India. There were Cretaceous dinosaurs there related to Cretaceous dinosaurs in other parts of the world. There were Pleistocene mammals in central India—bears, proboscideans, rhinoceroses, horses, pigs, hippos, deer, and cattle—that indicate definite connections with neighboring lands. But as for the span from Paleocene through Pliocene times, the record on the peninsula is blank. All of the Tertiary mammalian faunas of India, and they are abundant, are found along the borders of the Himalaya range, in the Siwalik Hills, and in Burma, regions that were part of the Laurasian land mass during past geologic history.

Perhaps the lack of Tertiary mammalian faunal evidence from the peninsular region of India is significant. Perhaps this was a drifting island continent, isolated from the rest of the world. But perhaps our lack of evidence is merely a matter of the accidents of preservation and possibly of discovery and collecting. Here, as in the case of Antarctica, future work in the field may lift a curtain that presently obscures the evidence that we seek for answers to the questions now looming large in our minds.

SUMMATION

Such is the history of the interrelationships of continents and land-living vertebrates, primarily mammals, in a large portion of the southern hemisphere during Cenozoic time, more than 65 million to a few thousand years ago. As contrasted with the history of mammalian faunas in North America, Eurasia, and Africa—a history of connected land masses, of Laurasia joined with the central, African part of Gondwanaland—this history of continents and land-living mammals below the equator is a story of isolated lands. It is a story of the several fragments of Gondwanaland, drifting toward their present positions with the independent evolution upon them of faunas separated from faunas in other parts of the world.

When one of these drifting island continents, South America, was reunited with the world to the north of it, the mammals that had for so long held sway on this southern land mass were overwhelmed by the influx of vigorous invaders from the north, so that by the end of Cenozoic time the faunal development of South America had in many respects become integrated with the faunal development of the continent at the northern end of the isthmian bridge. The other island, Australia, for which we have adequate paleontological and biological knowledge, remained inviolate from outside encroachments until the coming of modern man. And so its fauna remained unique.

With the advent of the Pleistocene epoch—the last great Ice Age—most of the continents were joined as they had not been joined for many millions of years, by land bridges rather than by broad continental ligations. Such connecting bridges affected the distributions of land-living vertebrates, with the resulting establishment in large part of the final patterns that distinguish the zoogeographical regions of today. To the story of the great Ice Age, its lands and its mammals, we will now turn our attention.

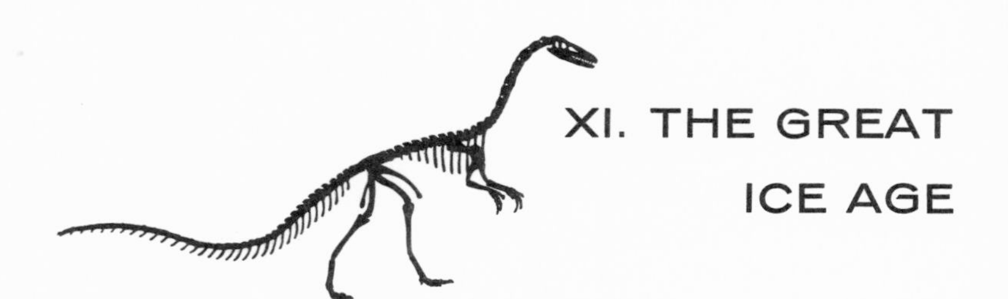

XI. THE GREAT ICE AGE

A COOLING WORLD

As the continents drifted toward their present positions during Cenozoic time certain developments took place, destined to affect very markedly the evolution and distribution of the land-living vertebrates inhabiting these lands. One such was the gradual lowering of world temperatures and the establishment of ever more distinct climatic zones during the progress of Cenozoic history. The Cretaceous world was of tropical and subtropical aspect over much of the extent of the continents—a world environment favorable to the wide deployment of gigantic dinosaurs. With the transition from Cretaceous into Cenozoic conditions there would seem to have been some lowering of temperatures, although not of sufficient magnitude to change markedly the environments in which the early Cenozoic mammals lived. These mammals lived on an earth that was still largely tropical and subtropical. But as time went on, the average decrease of world temperatures became ever more pronounced, with the gradual establishment of temperate and even cool climates in the higher latitudes. The banded climatic zones of the earth were coming into being. And, of course, environments within those continental blocks drifting across latitudinal zones, such as Australia and peninsular India toward the north, and Antarctica toward the south, were consequently affected. Needless to say, terrestrial life on such drifting land masses also was affected.

Another long-term development occurring simultaneously with the cooling of world temperatures was the elevation of the modern mountain ranges of the world—the Alps, Himalayas, Rockies, and Andes. This was the Laramide Revolution, having its beginning at the end of Mesozoic times and continuing through the early part of the Cenozoic. The mountain ranges—probably low at the beginning of Cenozoic history—reached higher into the skies as time went on, to form snow-capped barriers that affected local climates and variously restricted the movements of land-living vertebrates. The rise of mountains has occurred time and again through the long span of geologic history, and the building of those modern mountain chains, which so dominate landscapes in some of the continental areas, the result in part of forces within the earth causing vertical movements, and in part of the wrinkling of the crust along the leading edges of the drifting crustal plates, is only the last of numerous, recurring mountain-making events.

As a result of these events the Pleistocene world came into being—a world of extreme contrasts, with great polar ice caps (very likely the result of the North Pole being confined within a closed ocean basin and the South Pole being centered within a high, extensive continent), with temperature zones ranging from the Arctic and Antarctic climates of the poles to the tropical climates of the equator, and with enormous mountain ranges crossing the surfaces of the several continents. It was the world that we have inherited—with some changes and ameliorations, it is true—yet it was a world that is easily recognizable. It was the world in which our distant ancestors had their origins and in which they evolved. It was an atypical world, because during much of geologic history polar ice caps and strongly zoned climates evidently did not exist. It was a world that marked the culmination of a long series of events stretching back to the beginning of Cenozoic history.

BEGINNING OF THE ICE AGE

The definition and the limits of the last great Ice Age—known as the Pleistocene epoch—are not easy to establish, in part because we are so close to it. The history of the Pliocene merges into that of the Pleistocene, and even after a century or more of intensive study there is still not universal agreement as to where the lower limits of the Pleistocene

FIGURE 90. A drawing of a wild horse by a European paleolithic artist. The horses of the genus *Equus*, heralding the advent of the Pleistocene, were of this general type—small and stocky, commonly with rather large heads.

should be drawn. As for the upper limits of the epoch, one may say that we are still living in the Pleistocene. Of course, one may be arbitrary and call our age the "Holocene," or Recent epoch of geologic history, but in the large sense our time very possibly may be an interglacial stage within a sequence of recurrent glacial advances.

It is the glaciation of the northern hemisphere that particularly distinguishes the Pleistocene epoch from the preceding geologic epochs. Indeed, geologic practice for many years generally equated the Pleistocene with the successive advances and intervening retreats of immense ice caps from the north, periodically to cover vast expanses of North America and Eurasia with enormous continental glaciers. In recent years, however, there has been much sentiment for establishing the beginning of Pleistocene history at a point antedating by a considerable time span the advance from the north of the first continental glacier. This proposed beginning of Pleistocene history is conveniently marked by the almost worldwide spread of certain advanced mammals, particularly the modern horse, *Equus*, the early mammoths—cousins of our living elephants—and modern cattle, more or less typified by the genus *Bos*. These, and many others, are the animals characteristic of the so-called Villafranchian or Calabrian stage in Europe, of the Nihowan stage in China, of the Blancan stage in North America, and of the Kageran stage in Africa. The beginning of the Pleistocene, as represented by these age stages, can be placed at something on the order of three million years ago—perhaps a bit more. Its delineation on the basis of being recognizable throughout much of the earth, often far removed of the appearance and spread of modern mammals has the advantage from the direct effects of the northern glaciations.

In spite of its relatively short duration, as compared with other geologic epochs and periods, the Pleistocene is a large subject that has been intensively studied. Much could be said about the Pleistocene epoch and its life. Our interest, however, is concerned primarily with continual relationships during this last geological epoch and the effect of these, as well as of world climates, upon the distributions of land-living animals, specifically mammals. But before becoming involved with the problem of lands, intercontinental bridges, and wandering animals of Pleistocene time, it may be well to present the general chronology of the Pleistocene epoch as seen in Holarctica and Africa.

ICE AGE CHRONOLOGY

This chronology may be expressed in tabular form, as seen in the accompanying chart. As usual, according to geological practice, the table is arranged with the oldest divisions at the bottom and the youngest at the top, in accordance with the fact that sequences of deposits with their contained fossils range from the lower and older to the higher and younger.

	NORTH AMERICA	EURASIA	AFRICA
LATE PLEISTOCENE	Wisconsin glaciation	Würm glaciation	Gamblian pluvial
	Sangamon interglacial	Eem interglacial	Interpluvial
MIDDLE PLEISTOCENE	Illinoian glacial	Riss glacial	Kanjeran pluvial
	Yarmouth interglacial	Holstein interglacial	
	Kansan glacial	Mindel glacial	Interpluvial
	Aftonian interglacial	Cromer interglacial	
	Nebraskan glacial	Günz glacial	
EARLY PLEISTOCENE		Villafranchian (Europe)	Kamasian pluvial
	Blancan		Interpluvial
		Nihowan (China)	Kageran pluvial

FIGURE 91. Skeleton of *Elephas meridionalis,* an elephant, or mammoth, of early Pleistocene age.

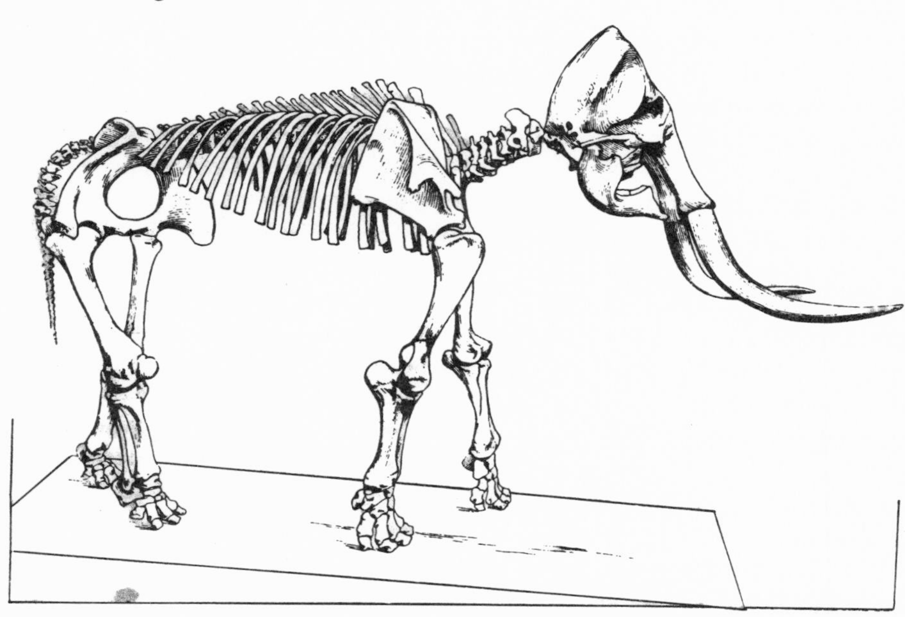

FIGURE 92. A restoration of *Elephas meridionalis*. The appearance and intercontinental migrations of elephants of this type can be used as criteria for the beginning of the Pleistocene epoch.

As will be seen, the early Pleistocene, which began perhaps three million years ago, is represented by the Blancan stage in North America (named from sediments containing very early Pleistocene fossils in Texas), by the more or less correlative Nihowan and Villafranchian stages, named from localities in China and Europe respectively, and by the Kageran and Kamasian pluvial, or wet, stages in Africa, these being separated by an interpluvial, or dry, stage. These several stages of early Pleistocene history, which lasted for about two million years, preceded the onset of the great glaciations in the northern hemisphere that so characterized the latter part of Pleistocene history.

There were four glaciations in North America and Eurasia, extending through middle and late Pleistocene time: the classic Günz, Mindel, Riss, and Würm in Europe and the Nebraskan, Kansan, Illinoian, and Wisconsin in North America. Each of these glacial stages represents a span of many thousands of years, during which immense continental glaciers advanced from the north, to cover the middle latitudes of the continents and to introduce Arctic conditions into lands that are now temperate.

Following each glacial advance there was a retreat—a so-called interglacial stage—when warm climates advanced to the north. We may very well be living in a fourth interglacial stage (which we call "Recent times"), following the Wisconsin and Würm glaciations.

Africa, being far removed from the continental glaciers, did not experience the alternation of prolonged Arctic and temperate conditions that so affected the northern hemisphere. On this continent, however, as well as in other parts of the southern hemisphere, as in Australia, for example, there were in many regions alternations of wet and dry intervals during Pleistocene time.

MODERN CONTINENTS AND BRIDGES

With the advent of Villafranchian time the arrangement of the continents assumes a familiar pattern, inherited from the continental relationships established during late Pliocene history and essentially similar to the distribution of modern continents. Briefly, Eurasia and North America were joined by the trans-Bering bridge, to form one great circumpolar, Holarctic land mass, extending through some 300 degrees of longitude, while South America was connected with North America by the Panamanian isthmus. India was by now an integral part of Eurasia, but separated from the land to the north by the massive

Himalayan range, which constituted a formidable barrier, even though it was not so high then as it is now. Africa was connected with Eurasia, much as it is today, through the Middle East. Australia and Antarctica were island continents, isolated from the rest of the world. Except for Australia and Antarctica, for a number of large islands such as Madagascar, New Zealand, and Greenland, and, of course, the numerous isolated small islands which during successive geologic ages and at various places have dotted the seas, the Villafranchian world was a world of continents joined by bridges, with lines of intermigration between Eurasia and North America, between Eurasia and Africa, and between North America and South America. It was through this arrangement of continents connected by bridges that numerous mammals moved back and forth, with modern horses, elephants, and cattle (using this word in a broad sense) leading the migrations.

GLACIERS AND THE MIGRATIONS OF NORTHERN MAMMALS

Following the long interval of Villafranchian time—a span of perhaps two million years during which the continents were covered by steppes, grasslands, and forest—the first of the great continental glaciers advanced over northern Eurasia and North America. As has been men-

FIGURE 93. Left. Skull of the Etruscan Ox (*Leptobos etruscus*) from the Villafranchian beds of northern Italy. This, one of the earliest of true cattle and rather similar to a large antelope, is, like the horse and the early mammoth, an indication of the onset of Pleistocene time. Right. Skull and jaws of the European Wisent *Bison priscus*. Bison invaded North America from Eurasia in the early Pleistocene.

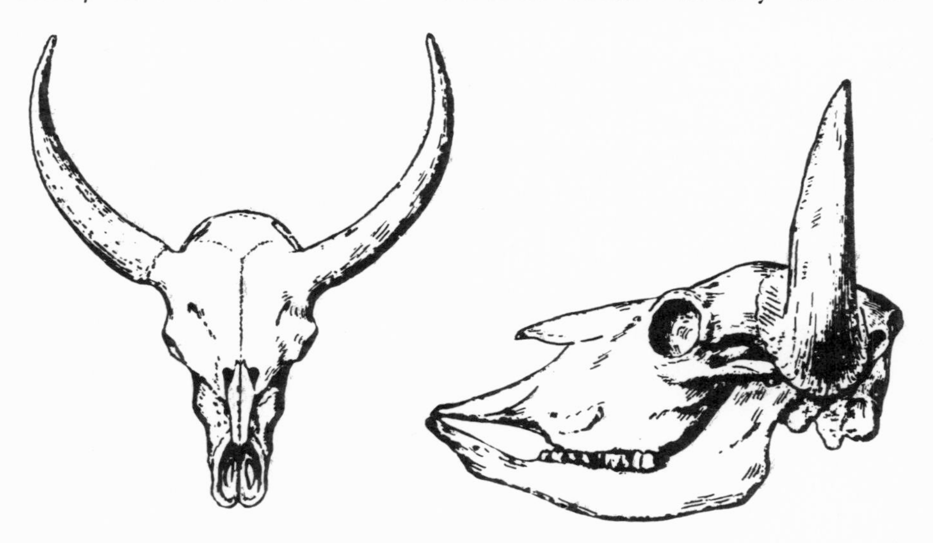

tioned, the glaciers came down four times to cover millions of square miles of land area—as far south as New York City and the southern borders of the Great Lakes in North America, so far south as to engulf Scandinavia and England and northern Europe, as well as portions of Siberia, in Eurasia. And south of the glacial fronts vast areas that today are smiling farmlands and centers of population were then Arctic tundra. Four times the ice retreated, so that the tundra lands and even considerable portions of the continents which had been ice-covered became forested and carpeted by grasslands, growing in relatively mild climates. Needless to say, these advances and retreats of the ice, with the effects of their movements reaching far beyond the glacial borders, strongly influenced the distributions of mammalian faunas. So we see two separate Pleistocene developments that partially controlled the manner in which animals were distributed across the lands: the connections between continents, which, of course, had been factors in previous geological epochs, and the advances and retreats of the continental ice sheets, which were something new in the history of mammals. Let us first examine the differing relationships of Holarctic faunas as influenced by the migrations between Eurasia and North America across the trans-Bering bridge, and then let us look at the manner in which such faunas were affected by the coming and going of the ice.

It would seem probable that there was a period of much migration across the trans-Bering bridge during Villafranchian time and extending on into the first glacial stage. This migration was a continuation of Pliocene movements that had increased in tempo toward the end of that geologic epoch. The Villafranchian was marked by the emigration of horses of the modern genus *Equus* into Eurasia, and from thence into Africa, so that the Old World became the home of asses and horses in the Palearctic region and of asses and zebras in Africa. The horses were accompanied on their migration into the Old World, across the trans-Bering bridge, by camels, which entered Asia but never reached so far to the west or the south as did the horses. It would appear lynx also crossed the trans-Bering bridge into the Old World. In the opposite direction elephants, commonly designated as "mammoths," crossed into the New World, but perhaps at a somewhat later stage in the Villafranchian than the westward movement of horses, camels, and lynx. By the end of the Villafranchian, however, mammoths were widely distributed throughout North America, as they were in Eurasia. It seems probable, too, that an aberrant hyaenid came into North America with the early mammoths.

During the glacial stages that followed the Villafranchian interval sea levels were lowered because of the sequestration of water (in the form of ice) in the great northern ice caps. (Today more than 90 percent of the ice and snow in the world is concentrated in the Antarctic ice cap; if this massive ice cover were to melt, the sea level would rise some 180 feet all over the world.) Consequently it is probable that much of the migration across the trans-Bering bridge took place during the glacial stages when sea levels had been lowered and when land connections were thus widely exposed. Migrations may therefore have been restricted during the first interglacial stage, but were probably renewed as the second great glacier advanced over northern lands. At this time it would appear that black bears came into North America from Eurasia, as well as certain saber-toothed cats, ancestral to the great sabertooth *Smilodon*, so characteristic of North America during late Pleistocene time.

Then during the third glacial stage movement of mammals into North America from Eurasia was augmented by bison and by deer of the genus *Cervus* (the Old World stag, the so-called elk or wapiti of the New World), and in addition by many carnivores, among which were foxes and wolves, weasels and wolverines.

Finally, during the fourth glaciation there was a most abundant movement of animals between Eurasia and North America. Among the mammals coming from the Old World into the New World across the trans-Bering bridge were moose (or elk, as they are called in Europe), reindeer or caribou, sheep, muskoxen (which eventually became extinct in Eurasia), and saiga antelope. Strangely enough, *Bison*, which had come into North America at an earlier time, would seem to have crossed back into Eurasia during this late phase of Pleistocene history.

MAN INVADES THE NEW WORLD

Near the end of the last glacial stage Man crossed into the New World from his home in Eurasia. These early men in the New World were of Mongoloid stock—the ancestors of the so-called Indians of America. The first of these paleo-Indians entered the New World perhaps as much as 12,000 or 15,000 years ago, as indicated by radiocarbon studies. Certainly they came into North America well before the extinction of various large mammals, specifically horses, camels, ground sloths, saber-toothed cats, mastodonts, and mammoths, as is shown not only by radiocarbon dating, but also by the undoubted association of artifacts, made by these men, with the now-extinct mammals.

It seems probable that the invasion of the New World by early man was not a simple and single crossing of the trans-Bering bridge; rather, these men would seem to have come in waves, one migration following another, through a considerable period of time. Very probably some of the first invaders crossed from Siberia to Alaska on dry land, before the formation of the present Bering Strait. Perhaps some of them crossed on the ice, and very probably some of them made the journey in skin-covered boats. Today, or at least in the very recent past, certain groups of Eskimos commonly and rather routinely crossed the strait back and forth between Asia and Alaska; their domain embraced the opposite tips of the Old and New World.

The evidence shows that some of the earliest immigrants pushed southward from Alaska into the northwestern portion of North America along dry valleys between great glacial tongues. Following them, other immigrants made their way south and extended their journeys through the continents by overland routes. It is likely that many journeys were also made along the coasts by boats.

Moreover, the ancient Mongoloids may not have been the only invaders of the Americas. For example, there is some evidence to indicate that on occasion Polynesian peoples drifted onto the western coasts of the New World, where they either perished or were absorbed by the peoples whom they met. As a final note, it should be mentioned that some stelae found in Middle America portray the likenesses of long-nosed, heavily bearded men, which raises the possibility that Europeans may have reached the American shores long before the Vikings. But we are here becoming involved with a time that may be considered as of "post-Pleistocene" age and thus somewhat beyond the proper limits of this discussion.

The point should be emphasized, however, that Man was a part of the late Pleistocene invasions of the New World. And after he reached the two Americas he became established in harmony with the animals that surrounded him. Indeed, early Man was a part of the fauna, living in balance with it for many thousands of years.

INVADERS AND STAY-AT-HOMES

So we see there was a pronounced interchange of mammals between Eurasia and North America throughout the extent of Pleistocene history—an interchange that was in many respects more balanced, because it was less selective, than the interchanges of previous geologic epochs. The movements of mammals from Asia to North America were especially significant, because they introduced many of the large mammals that were to dominate North American faunas during Pleistocene and Recent history. Such immigrations into the New World established here various mammoths, which flourished in great numbers until a few thousand years ago, deer, sheep, "cattle" in the form of *Bison*, and many large carnivores. Horses and camels were indigenous to North America and lived on this continent throughout the Pleistocene. With the beginning of Pleistocene times these animals entered the Old World to continue there until the present day, although in North America they became extinct, as did the mammoths.

Yet even though these migrations were important, especially as they introduced large mammals into the New World, they involved comparatively small fractions of the faunas that inhabited the Palearctic and Nearctic regions. As in previous Cenozoic epochs, only a minor proprotion of the animals seemingly available for the crossing made the trip. Perhaps this was in large part owing to the high northern position of the trans-Bering bridge, which would have eliminated many animals from attempting the journey—not only because of the cold, but also because of the lack of food for which they were adapted. Obviously such animals as warmth-loving hippopotamuses, which ranged as far north as England during interglacial stages, would not wander across a trans-Bering bridge. Nor would pigs, since these are forest animals and would be deterred by large expanses of tundra. Likewise, pronghorns in America might not make the crossing into the Old World because of the lack of suitable grasses on which to graze, even though sheep and bison did utilize the bridge.

But it is something of a puzzle as to why the woolly rhinoceros, an inhabitant of the Eurasian steppes, did not cross into North America. In Eurasia the woolly rhinoceros was almost a constant companion of the woolly mammoth; these two mammals loom large in the cave paintings of early Man in Europe. The woolly mammoth came into North America, where it lived abundantly, but the woolly rhinoceros failed to make the crossing.

While various mammals were passing between Eurasia and North America by way of the Pleistocene trans-Bering bridge, other mammals were making their separate ways into North America from South America, and conversely into South America from the north, by way of

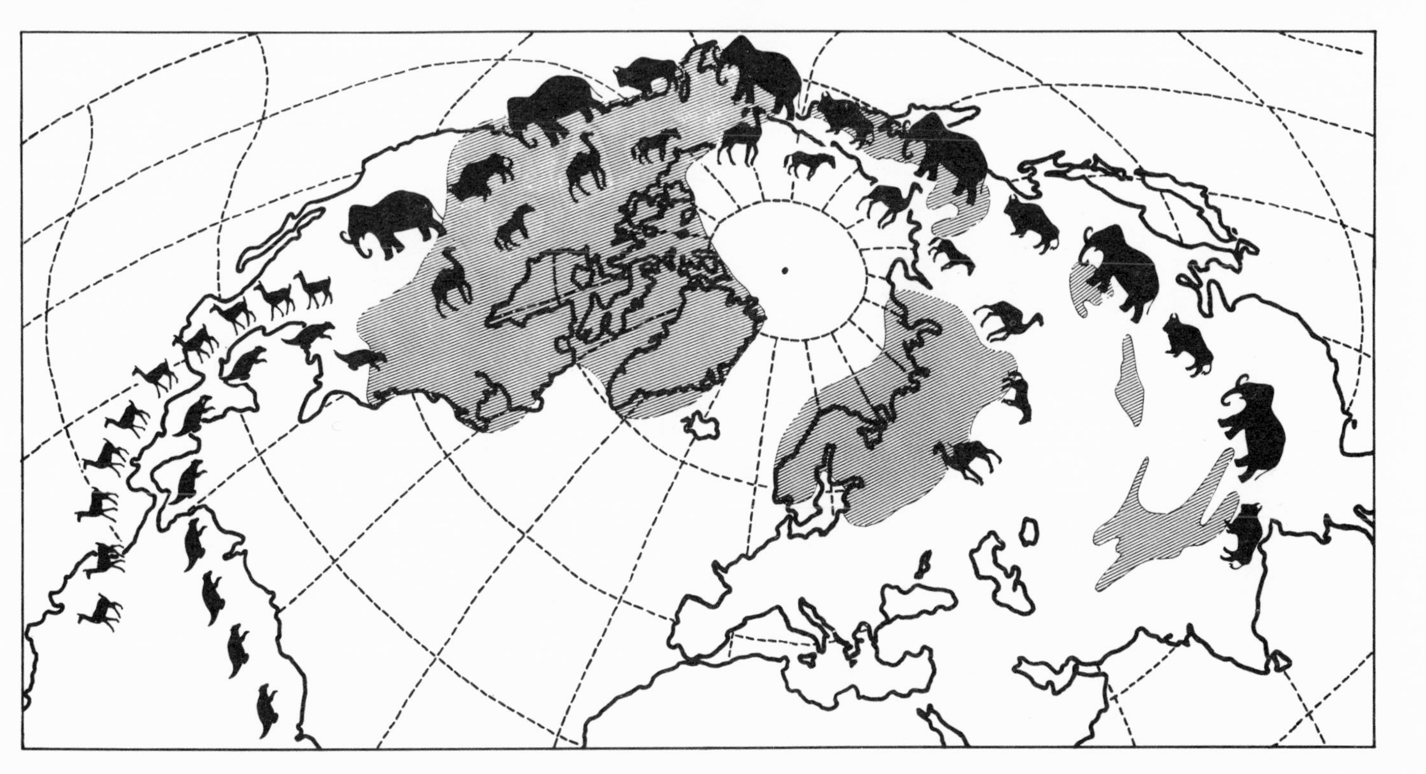

FIGURE 94. The advent of Pleistocene time was marked the wide intercontinental migrations of certain large mammals. Modern horses of the genus *Equus* crossed from North America, their place of origin, by way of a trans-Bering bridge, to spread throughout the Old World. At the same time early mammoths and bison crossed by the same path into North America. Even before the beginning of Pleistocene history there were movements of mammals up and down the newly established Panamanian bridge. Small camels (llamas) pushed into South America from their North American center of origin, while gigantic ground sloths moved from South into North America. Yet some mammals were curiously limited. The woolly rhinoceros, so common in Eurasia, never reached North America, and the American pronghorn antelope never entered Eurasia.

the Panamanian isthmus. These movements have already been outlined in the last chapter, so they need not be described in detail here. Suffice it to repeat that gigantic ground sloths and armored glyptodonts invaded North America from the south, accompanied by armadillos, capybaras, and porcupines, these last perhaps being the latest of the immigrants, to augment the developing Nearctic faunas of the Pleistocene. And as we have seen, there was a rather massive invasion of North American mammals into South America along the Panamanian bridge. This influx of many carnivores and herbivores inundated most of the long-established native mammals of South America, to form the Pleistocene and Recent neotropical mammalian faunas, as we know them. Of course, the Panamanian bridge was not directly affected by the advances and retreats of the northern glaciers, so it is probable that movement of mammals along its length was more continuous and less interrupted than the intercontinental migrations between eastern Asia and northwestern North America.

A partial summary of the animals that utilized the trans-Bering and the Panamanian bridges during the extent of Pleistocene time may be presented as follows:

EURASIA		NORTH AMERICA		SOUTH AMERICA
	←	Opossums	→	
	←	Beavers		
			←	Capybaras
			←	Porcupines
			←	Armadillos
			←	Glyptodonts
			←	Ground sloths
	←	Canids	→	
Bears	→		→	
	←	Raccoons	→	
	↔	Cats	→	
	←	Horses	→	
		Tapirs	→	
		Peccaries	→	
		Llamas	→	
	←	Camels		
Deer	→		→	
Sheep	→			
Muskoxen	→			
Saiga	→			
Bison	→			
Mastodonts	→		→	
Mammoths	→			
Man	→		→	

ORIGINS OF THE MODERN AFRICAN FAUNAS

During the years of the Pleistocene, when animals were migrating back and forth between Eurasia and North America and between North America and South America, there were also intercontinental migrations between Eurasia and Africa. It would seem, however, that the movement of mammals north and south, and especially in the southerly direction from Eurasia to Africa, was not so extensive as has been hitherto supposed. For instance, it was long thought that the Pleistocene and modern African faunas were largely derived from Eurasia—that there was a southward migration of mammals, especially large mammals, and in particular the browsing and grazing hoofed mammals, as exemplified by the antelopes, giraffes, and cattle. It was supposed that these had developed largely in the Eurasiatic region. Recent studies indicate, however, that such probably was not the case—at least to any large extent.

The reasons for this later, modified view of African faunal relationships are found in the nature of Villafranchian mammals discovered in Africa during the past few years. In short, the earliest African mammalian faunas are predominantly African in character—indicating a considerable period of evolutionary radiation on the African continent *before* the advent of Pleistocene time. The details of this inferred pre-Pleistocene development of mammalian faunas in Africa cannot be seen very clearly, because Pliocene mammalian faunas in Africa are of limited extent. Nevertheless there are some clues, and certainly by the beginning of the Pleistocene it is evident that Africa was the home of a large and varied indigenous fauna. In the Villafranchian deposits of Africa are found large cats, including saber-toothed cats, and hyenas, horses of the *Hipparion* type holding over from the Pliocene, as well as the modern horse genus *Equus*, rhinoceroses closely related to the present-day white rhinoceros, a considerable variety of pigs, including a gigantic warthog, hippopotamuses, a very large, heavy, oxlike giraffe, *Libytherium*, with spreading horns, gazelles and cattle, the peculiar dinothere proboscideans and mastodonts, left over from Pliocene times (and by the Villafranchian in the final stages of their evolutionary histories), early mammoths, baboons, including a gigantic form, and early ape-men—the men designated as *Paranthropus* and *Australopithecus*. Although there are evidences in this listing of some communication with Eurasia, especially as indicated by such forms as saber-toothed cats, horses, gazelles, and proboscideans such as dinotheres and mastodonts, the bulk of the representation has a distinct African character. Africa obviously was a center for the evolution of many of the mammals that still are typical of this great continent.

Indeed, it is quite as valid to think that many of the resemblances between Africa and Eurasia, as displayed in Pleistocene mammals, may be the result of migration out of Africa toward the north, rather than into Africa from the north, as formerly supposed. Thus it seems probable that certain civets, hyaenids, and cats migrated out of Africa into Eurasia during the transition from Pliocene to Pleistocene time, that the aardvarks made such an emigration at an even earlier date, as did numerous proboscideans and rhinoceroses, and that Africa was a center from which pigs, hippopotamuses, tragulids, giraffes, and antelopes moved northward, in Pliocene and early Pleistocene times.

Africa's particular importance, especially to our anthropocentric eyes, was as a center for the evolution of monkeys and apes, and in particular those ape-men that were on the line of evolutionary development leading to early men. The characteristic Old World monkeys appear to have arisen in Africa and to have migrated into Eurasia during middle Tertiary times, well before the opening of the Pleistocene. Likewise, upon the basis of fossil evidence, certain apes appear to have had African origins.

Some of the most significant of fossil discoveries have been of Villafranchian anthropoids, the ones mentioned above under the names of *Paranthropus* and *Australopithecus*, in South and East Africa. These fossils have been found in considerable numbers—at least for fossil primates—so that our knowledge concerning the characters of the basic human stock is founded upon a truly respectable body of evidence. Some of the discoveries have been made and studied in South Africa by Raymond Dart, Robert Broom, H. B. S. Cooke, Philip Tobias, and others, while some very spectacular finds have been made in recent years at Olduvai Gorge in East Africa by Louis Leakey and his associates. Olduvai Gorge is indeed a remarkable treasure-trove in the search for ancient man, and it is there, as a result of the fortunate deposition of ancient anthropoid bones in conjunction with the outflow of lavas, that it has been possible to establish radiometric dates for those ape-men known as the Australopithecines, and their associates, showing them to be much older than hitherto had been supposed.

Paranthropus was a large and very robust anthropoid, standing erect and searching for his food among the vegetation surrounding him. *Australopithecus* was a smaller ape-man contemporaneous with *Paran-*

thropus, also erect in stature, but seemingly more of a hunter—pursuing animals and killing them very possibly with crude weapons. From such beginnings, especially as represented by *Australopithecus,* early men evolved, to migrate into Eurasia, where they appear as the well-known *Pithecanthropus* in early post-Villafranchian time.

So we see that the Villafranchian interval was the time when modern African faunas took form. And we have noted that these faunas were to a large degree indigenous, that they evolved along definite African lines, from which various emigrants made their way north into the Eurasian land mass. And not the least of these emigrants were the early men, derived from a basic African stock.

Yet even though there were migrations out of and into Africa during Villafranchian time, the movements of mammals north and south were not of large volume, as compared with the faunas then living in Africa and Eurasia. Here we see a parallel to the comparatively limited intercontinental migrations across the trans-Bering bridge. If there had been extensive intercontinental migrations, one would expect to find more mixing of faunal elements in Eurasia and Africa. A large degree of mixing failed to take place; the Villafranchian faunas of Eurasia and of Africa had their distinctive characters, largely inherited from preceding Pliocene assemblages.

All in all, the development of Pleistocene land-living mammals in Africa was largely a separate chapter in the evolution of vertebrate faunas on the several continents. In spite of connections that existed, there was even less coming and going between Africa and Eurasia than there was between Eurasia and North America. And for this reason the fauna of the Ethiopian zoogeographical region today is quite distinct, contrasting sharply with the faunas of the adjacent Oriental and Palearctic regions. It makes one realize that, although connections between continents may be important in the distributions of land-living animals, they do not necessarily bring about uniformity among the faunas inhabiting the several continental areas.

GLACIAL AND INTERGLACIAL FAUNAS

Some mention has been made of the successive advances and retreats of the great continental glaciers across the northern lands of Holarctica during much of Pleistocene time. And as we have seen, they affected the intercontinental migrations of northern mammals, especially across the trans-Bering bridge—a bridge that was particularly available during periods of glacial advance, when ocean levels were lowered by the concentration of water in the great glaciers, thus baring land surfaces that otherwise might have been under water. But the effects of the great ice sheets were not limited to the east and west movements of animals across the high-latitude Bering bridge; there were inevitable correlations between north and south movements of the ice and the north and south distributions of Pleistocene mammals in Holarctica.

Four times the continental glaciers advanced across northern North America and northern Europe, obliterating forests and glades, and, as has been noted, causing the lands in front of the glaciers to be transformed from areas of abundant vegetation to regions of tundra and stunted trees. And four times, as the ice sheets advanced, the land-living animals retreated southward—with warmth-loving species occupying habitats far to the south of the regions in which they had formerly lived, and with cold-adapted species taking over the regions from which the former inhabitants had been so forcibly driven. The results of these southward migrations of animals are exhibited by fossil remains found in the northern hemisphere, in sediments contemporaneous with the glacial advances. Moreover, most fascinating and lively pictures of the animals that inhabited Europe during the last glacial advance were painted and incised on the walls of caves that had been frequented by our Paleolithic ancestors.

At these times central and southern Europe was inhabited by wolves and foxes, gigantic cave bears, lynx and other cold-climate carnivores, small, shaggy horses, woolly rhinoceroses, forest pigs, reindeer, bison, and the wide-ranging woolly mammoths. These animals we know from their bones and (so far as the last glaciation is concerned) from pictures painted on the walls of caves, as has been mentioned. In addition, however, carcasses of some of these animals, especially the woolly rhinoceros and mammoth, have been found intact—rhinoceros bodies in petroleum deposits in Rumania, mammoths frozen in the ice in Siberia. Thus we have a graphic impression of what life was during such cold spells in Europe.

Similar incursions of mammals adapted to cold climates occurred across central North America. The fossil remains of many of the mammals frequenting the tundras of central Europe are found in what were the tundra areas of North America as well. Here one finds various carnivorous mammals, horses, but not rhinoceroses, no pigs, but reindeer and bison, and in addition muskoxen, woolly mammoths, and mastodons. As these mammals ranged across the hills and plains in front

FIGURE 95. Europe during the last Ice Age, a time when great glaciers advanced far into present temperate regions, when winters were long and snows were deep. A Paleolithic man and child are trying to hide from their archenemy the huge cave bear (*Ursus spelaeus*). In the background are woolly mammoths (*Mammonteus primigenius*).

FIGURE 96. The woolly mammoth, which roamed across the northern hemisphere in late Pleistocene time, disappeared from Eurasia and North America some thousands of years ago. This great, hairy elephant was well known to the men who inhabited northern lands, and numerous pictures of the mammoths were produced by the so-called cave men of Europe.

of the glaciers, the mammals that had previously occupied such regions found refuge in lands to the south—in the northern portions of Middle America and in the Florida peninsula. Florida must have been a populous haven for the animals that could not adapt themselves to the low temperatures near the ice fronts. Here were many of the carnivores that had moved south, including saber-toothed cats, the ground sloths, glyptodonts, and the capybara, that had come into North America by way of the Panamanian isthmus, and numerous horses, tapirs, peccaries, deer, bison, and the large southern mammoths (not the northern woolly mammoth) that shared this habitat with the American mastodon.

During the interglacial stages, both in Eurasia and in North America, there were, of course, northward movements of faunas as the glaciers retreated. So it was that we find such animals as lions, hyenas, and hippopotamuses inhabiting the interglacial lowlands and swamps of southern England, and correspondingly, giant beavers, tapirs, sloths, and mastodonts in the interglacial landscapes of New Jersey, Pennsylvania, and other northern states.

So the years of the Pleistocene flowed into Recent time, which may be, as has been mentioned, an interglacial interval, and thus still within the limits of the Pleistocene epoch. Whether we choose to think of ourselves as living in Recent or Pleistocene time, it is a fact that the northward movement of animals, following the retreat of the last glacier, would seem to be still in progress. Within historic time in North America such typically southern mammals as the Virginia opossum (*Didelphis*), the armadillo (*Dasypus*), and the ring-tailed "cat" (a small raccoon, *Bassariscus*) have been moving northward toward the northern states and Canada. And the same applies to various birds, such as the cardinal and the tufted titmouse. The distributions of animals are constantly being modified, and it is interesting to see changes, such as the ones mentioned above, taking place within the course of a century or less, evidently the result of natural changes in climates, small though these may be.

PLEISTOCENE EXTINCTIONS AND MAN

About 10,000 years ago, or less, there were some rather dramatic changes in the faunas of North America and Eurasia, as well as those in some other parts of the world. These were the selective extinctions of various large animals that had been characteristic of Pleistocene times. In Europe the gigantic cave bear, which had loomed large in the life of Paleolithic men, died out, as did saber-toothed cats, the mammoths, the woolly rhinoceros and other rhinoceroses then inhabitants of Eurasia, certain species of bison, and various rodents. Other mammals long typical of the European scene retreated from these latitudes to continue elsewhere. Such were the hyenas, leopards, lions, wild dogs, and hippopotamuses that moved south, and the saiga antelope, muskoxen, and Arctic hares that moved north. At the same time, in North America, the great saber-toothed cats, which had flourished for so long, became extinct, as did the gigantic ground sloths and glyptodonts, the American mastodonts and the mammoths, horses, camels, certain types of pronghorns, muskoxen, and some bison, these last characterized by enormously long horns. And as in Eurasia, some of these mammals, notably the horses and camels, although becoming extinct in North America, continued their existence elsewhere—in the Old World.

Similarly, various large mammals became extinct in South America at about this same time—notably ground sloths and glyptodonts, horses and mastodonts. In Africa there were large mammals, mostly related to mammals still living on that continent in late Pleistocene times, that disappeared. The same is true even for so isolated a continent as Australia. Gigantic wombats, known as "diprotodonts"—as large as grizzly bears—inhabited Australia in subrecent times, evidently becoming extinct only a few thousand years ago. And with them were giant kangaroos, also no longer living.

The line of demarcation between the Pleistocene and so-called Recent times frequently has been set at the retreat of the last glaciers. It has also been placed at the time of extinction of large mammals—which has the convenience of being a widely distributed phenomenon.

Whether the Pleistocene has ended or not, one may ask why there was such a broad disappearance of large mammals during the interval perhaps between about 10,000 and 20,000 years ago. Could such a selective extinction have been caused entirely by climatic changes or other natural causes? And if one is to attribute the disappearance of the woolly mammoth and the woolly rhinoceros, for ex-

FIGURE 97. The extinctions of various large mammals in North America.

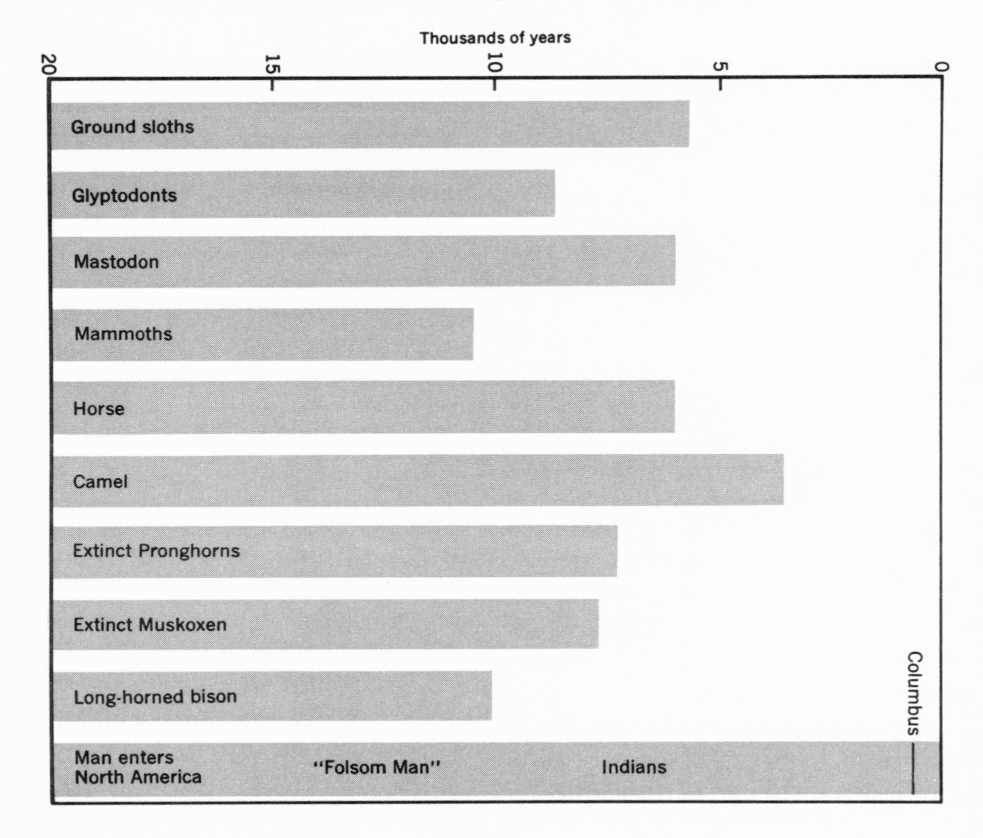

ample, to the retreat of the glaciers and the expansion of warm temperatures across their former environments, how is one to explain the disappearance at the same time of ground sloths in South America or diprotodonts in Australia—where the factors of glacial retreats were not involved? As has been noted, there probably were extensive temperature changes, correlated with the northern glaciations, even in continental regions far removed from glaciers. Such changes would have brought about sufficient differences in vegetation or other ecological factors within the environment to cause the demise of the large mammals.

However, the question has been raised as to whether early Man had anything to do with the extinction of the large mammals at the "end" of Pleistocene time. In Eurasia and Africa Man had been living with the mammals of the Pleistocene for hundreds of thousands of years, and presumably was well adjusted to them and to his habitat. In brief, Man in the Old World was an integral part of the fauna. So why should he have had any effect on the large mammals of very late Pleistocene time?

One should consider that during the final stages of Pleistocene history, notably during the third interglacial and fourth glacial stages, there was a progressively increasing development of human cultures. There was, in effect, an "explosion" of the mind, with a consequent rapid improvement of tools and technology. Neanderthal man, a skillful toolmaker, was superseded and displaced by Cro-Magnon man, who was in many respects an even more skillful toolmaker and almost certainly a more inventive and a more highly organized social person than his Neanderthaloid predecessor. Perhaps it was the development of better weapons and better methods of hunting, combined with small climatic changes, that had such a profound impact upon the large Pleistocene mammals of Eurasia. Yet one still wonders how these early men, certainly living in small populations, could have so efficiently brought about the end of such a considerable number of large, frequently aggressive mammals.

We have seen that the early Mongoloid invaders of the New World reached their new domain before the extinction of many large mammals, such as ground sloths, mammoths, mastodonts, horses, and camels. So one is tempted to speculate that perhaps in the New World the entrance of man into a delicately balanced ecological situation had its devastating effects. Again, however, one wonders how admittedly small populations of hunters, armed with primitive weapons, could have caused the extinction of the North American horses and mastodonts, which would appear to have inhabited the continent in great numbers, if the frequency of their fossil remains forms any sort of a criterion of abundance.

These are questions that have been vigorously debated in the past and will continue to be debated in the future. Whether Man was directly responsible for the extinctions of large mammals at the so-called end of the Pleistocene, the fact is that by this stage of geologic history he had become a mobile and perhaps a disturbing influence in animal associations all over the world.

So our story of the distribution of land-living animals in relation to the arrangements of the continents draws to its close with the appearance of Man as a powerful factor in determining the ranges and even the very presence or absence of animals living with him. The beginnings of the disturbing influences of Man among the animals around him are seen throughout the world during the closing years of the Pleistocene. With the coming of what we call "Recent" times the influence of Man on the life of the continents becomes increasingly disturbing and finally, and all too frequently, disastrous. To this unhappy situation we will now briefly turn our attention.

XII. MAN, MAMMALS, AND CONTINENTS

THE IMPACT OF MAN

Man was a disturbing influence on the life of the continents, even when he was a primitive hunter armed with comparatively crude weapons. By the time he had attained the Neolithic stage of cultural development, armed with polished stone weapons and tools, with bone and wooden artifacts, and with an increasingly sophisticated social organization, his impact upon the life around him was correspondingly enlarged. Subsequently, in the early metal-working cultures, of bronze and later of iron, the pressure of Man on his environment was still increased. The process has been continuing ever since.

As we have seen, it is very possible that ancient Man was instrumental in the extinction of at least some of those large mammals so characteristic of the late Pleistocene world. And as Man entered upon the Neolithic and the later metal-age stages of his development he probably still caused disappearances of animals—sometimes in a direct fashion, and again in more indirect ways. One of the dramatic ways in which Man changed the life around him was through the process of domestication.

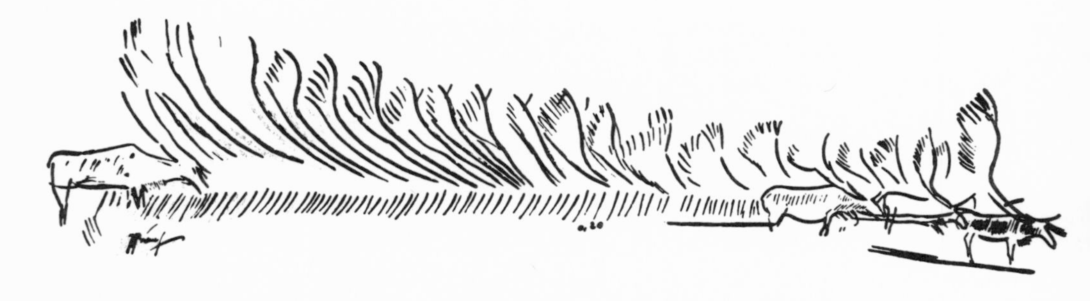

FIGURE 98. Man was a hunter long before he began to domesticate animals and to take them with him on his wanderings. A Paleolithic man in Europe made this impressionistic drawing of a reindeer herd many thousands of years ago. It was a herd to be viewed and pursued; only his relatively modern descendants learned how to domesticate the reindeer—and other useful animals.

MAN AND HIS SATELLITES—THE DOMESTIC ANIMALS

Man's first domesticated animal was the dog, derived from Old World wolves, to become his constant hunting companion. But as Man gradually turned from hunting to early practices of agriculture he domesticated various animals, along with plants, to broaden and ensure his food supply. Thus in the course of time he was the master of camels cattle, sheep, goats, pigs, asses, horses, fowl—and many other animals as well. At the same time Man often became more mobile than he had been as a primitive hunter, especially when he had adapted horses, camels, and oxen to the tasks of carrying and pulling him and his burdens. Consequently he moved about, taking with him the various animals that were his companions and servants—to modify animal distributions on the continents.

In Eurasia and Africa Man shifted his domestic animals about within continents where their ancestors had lived as wild beasts. And in so moving about, various animals often were taken into regions where neither they nor their ancestors had previously been present. Sometimes the introductions of domestic or partially domestic animals into areas which formerly had not known their presence affected some of the wild species living in these regions. Perhaps wild animals were restricted in their ranges because of the space occupied by Man and his domestic animals; perhaps they were even hunted down and exterminated, if they seemed to be dangerous because they preyed upon domestic flocks or if they seemed to compete with domestic animals for the available food supply. The effects of Man and his animal satellites upon the environment around him were varied and numerous. In essence, Man, even as an early hunter-agriculturist, modified the distributions of animals wherever he went.

The Mongoloid invaders of the New World brought with them only the dog, and with the dog they were content to live for many millennia. But as some of them developed their own civilizations they domesticated a few of the native wild animals—notably llamas in South America, and turkeys. The aboriginal invaders of Australia brought with them the dingo, and that was their only domestic animal, although now the dingo is commonly wild. The early settlers of the Pacific islands, particularly the Polynesians and Melanesians, often carried with them dogs, pigs, and chickens, and distributed these animals far and wide across Pacific expanses.

MAN'S MODIFICATIONS OF ANIMAL DISTRIBUTION

Consequently at an early stage in the evolution of human cultures Man had modified the distributions of animals, especially mammals, throughout the world. Yet although the effects of Man's activities with regard to the distribution of animals were numerous, they were until fairly recent times of comparatively minor consequence. Even as late as two millennia ago human populations were relatively limited and

FIGURE 99. By the third millennium B.C. the domestication of many animals had proceeded to a sophisticated level, as indicated by this beautiful depiction of a bull from Mohenjo-Daro, in the Indus Valley.

on the whole rather self-contained. True enough, there were migrations and invasions (such seem to have been inevitable in the course of human history), but their effects generally were not profound. Europe was the home of Caucasians and their domestic animals, and even though by the beginning of the Christian era the Roman Empire stretched from northern Britain to Persia and from the Pillars of Hercules into the Middle East, the inevitable movements of peoples and animals within this great domain were largely limited to Europeans and the animals under their control. Likewise, Africa south of the Sahara was the home of the Negroids, much of Asia of the Mongoloids, and the Americas—then quite unknown to the inhabitants of the Old World—of that branch of Mongoloids that we call "American Indians." And in Australia there were the aborigines. There was not much large-scale movement and mixing of populations, with the consequent shifting about of domestic animals.

MAN AND OTHER ANIMALS IN THE MODERN WORLD

The great age of exploration, having its inception with the coming of the sixteenth century, saw the beginnings of changes that were to affect peoples and animals on all of the continents. From then until the present day there have been progressively increasing movements of people, with correlative displacements of animals.

Vast numbers of cattle, sheep, pigs, horses, and other domestic animals now inhabit the New World, where 500 years ago they were unknown. At the same time, many of the animals that once were so characteristic of North and South America—bison, pronghorns, bears, cougars, wolves, peccaries, and others—are now restricted to mere remnants of their former numbers or are even on the verge of extermination. Australia has become a land of sheep and cattle, and plagued by rabbits, where once marsupials held sway. Likewise, New Zealand, once almost devoid of mammals, now harbors millions of grazing sheep and has taken on the appearance of another England. Even in Africa the once great hordes of varied antelopes, buffalo, giraffes, and their accompanying predators, the lions, leopards, hyenas, and such, are now largely restricted to game parks, protected for the delectation of sightseers. Their ranges are being increasingly occupied by domesticated animals.

Thus the zoogeographical regions of the world, as commonly delineated upon a world map, have become to a considerable degree remembered rather than real regions. The traveler may have to put forth much time and effort to obtain even partial glimpses of the Nearctic fauna in the United States, the Oriental fauna in India, or the Ethiopian fauna in South Africa. And if it were not for the setting aside and vigilant preservation of national parks, wildlife refuges, and game parks, even such glimpses might now be largely beyond the limits of possibility. Man has profoundly altered the surface of the earth within the past century or two, and in hardly any aspect has this alteration been more extreme than in the distributions of the original inhabitants of the several continents.

Man is determined that the land areas of the earth are for his exclusive use, and such use involves the raising of the many domestic animals from which a large part of his food supply is obtained. Any other animals, if they are "in the way," must be eliminated. Moreover, this vast disturbance of animal populations is not Man's only alteration of a world as it used to be. The changes affect human populations as well, as has already been mentioned. Man is ever more mobile, and with that mobility the distinctiveness of racial types has been and will continue to be ever more blurred. Thus the continents, originally the domains of comparatively pure races, are now lands where mixed populations are becoming ever more mixed.

Consequently the world is growing into a different world—as it never was before in the long span of evolutionary history. In essence it is becoming a zoologically impoverished world. We can study the distributions of animals upon the continents and the effects of continental relationships upon such distributions—as they were in the distant past and as they have been in the immediate past—but such studies are increasingly separated from the world in which we live by our propensity for meddling with and changing the ecology around us. Nevertheless, the key to the present and to the future is in the past, and it behooves us to continue these studies with all of our facilities and power, if we are to enjoy an understanding of the world so necessary for our very survival in the future.

BIBLIOGRAPHY

The literature bearing upon the content of this book is truly vast. Moreover, it is necessarily diffuse, being concerned with numerous scientific disciplines and philosophies, some of which at first glance would seem to have little in common with some others. There appears to be little common ground, for example, between geophysics and vertebrate paleontology, or between structural geology and herpetology. Yet these diverse fields and many more bear upon the subject of ancient continents and the distributions of land-living vertebrates through time.

Because of the volume and diversity of the literature, no attempt will be made at this place to list all or even a sizable fraction of the works consulted in the preparation of this volume. Some of the more important books and a few of the scientific papers bearing on the subject are cited.

It is to be hoped that the references thus set forth may be helpful.

American Philosophical Society. 1968. "Gondwanaland Revisited: New Evidence for Continental Drift." *Proceedings of the American Philosophical Society,* Vol. 112, No. 5, pp. 307–353. (A symposium, with articles by Gerard Piel, J. Tuzo Wilson, Marshall Kay, Frederick J. Vine, Alfred Sherwood Romer, and Walter M. Elsasser.)

AMOS, ARTERO J., ed. 1969. *Gondwana Stratigraphy.* IUGS Symposium, Buenos Aires, xvi+1173 pp.

ANDERSON, H. M., and ANDERSON, J. M. 1970. "A Preliminary Review of the Biostratigraphy of the Uppermost Permian, Triassic and Lowermost Jurassic of Gondwanaland." *Palaeontologica Africana,* Vol. 13, Supplement, pp. 1–22, Charts 1–22.

BULLARD, EDWARD. 1969. "The Origin of the Oceans." *Scientific American,* Vol. 221, pp. 66–75.

CALDER, NIGEL. 1972. *The Restless Earth: A Report on the New Geology.* New York: The Viking Press, 152 pp.

COLBERT, E. H. 1969. *Evolution of the Vertebrates.* 2d ed. New York: John Wiley & Sons, Inc., xvi+535 pp.

COX, C. B. 1967. "Changes in Terrestrial Vertebrate Faunas during the Mesozoic." *The Fossil Record,* pp. 77–89. Geological Society of London.

DARLINGTON, PHILIP J., JR. 1957. *Zoogeography: The Geographical Distribution of Animals.* New York: John Wiley & Sons, Inc., xi+675 pp.

DARWIN, CHARLES. 1880. *The Origin of Species by Means of Natural Selection.* 6th ed. London: John Murray, xxi+458 pp.

———. 1965. *Biogeography of the Southern End of the World.* Cambridge, Mass.: Harvard University Press, x+236 pp.

DIETZ, ROBERT S., and HOLDEN, JOHN C. 1970. "The Breakup of Pangaea." *Scientific American,* Vol. 223, pp. 30–41.

——— and SPROLL, WALTER P. 1970. "Fit between Africa and Antarctica: a Continental Drift Reconstruction." *Science,* Vol. 167, pp. 1612–1614.

DUNBAR, CARL O., and WAAGE, K. M. 1969. *Historical Geology.* 3rd ed. New York: John Wiley & Sons, Inc., 556 pp.

DU TOIT, A. L. 1957. *Our Wandering Continents.* Edinburgh: Oliver and Boyd, xiii+366 pp.

FOODEN, JACK. 1972. "Breakup of Pangaea and Isolation of Relict Mammals in Australia, South America, and Madagascar." *Science,* Vol. 175, pp. 894–898.

FUNNELL, B. M., and SMITH, A. GILBERT. 1968. "Opening of the Atlantic Ocean." *Nature,* Vol. 219, pp. 1328–1333.

GEORGE, WILMA. 1966. *Animal Geography.* London: Heinemann Educational Books, Ltd., x+142 pp.

GREEN, W. L. 1857. "The Causes of the Pyramidal Form of the Outlines of the Southern Extremities of the Great Continents and Peninsulas of the Globe." Edinburgh: *New Philosophical Journal,* Vol. 6, New Series.

HALLAM, ANTHONY. 1967. "The Bearing of Certain Palaeozoogeographic Data on Continental Drift." *Palaeogeography, Palaeoclimatology, Palaeoecology,* Vol. 3, pp. 201–241.

———. 1971. "Mesozoic Geology and the Opening of the North Atlantic." *Journal of Geology,* Vol. 79, pp. 129–157.

HAUGHTON, S. H., ed. 1972. *Second Gondwana Symposium, South Africa, 1970.* Pretoria: Council for Scientific and Industrial Research, v+689 pp.

HEIRTZLER, J. R. 1968. "Sea-Floor Spreading." *Scientific American,* Vol. 219, pp. 60–70.

HURLEY, PATRICK M. 1968. "The Confirmation of Continental Drift." *Scientific American,* Vol. 218, pp. 52–64.

———. 1971. "Possible Inclusion of Korea, Central and Western China, and India in Gondwanaland." American Geophysical Union Meetings, Washington, April 1971.

IRVING, E. 1964. *Paleomagnetism and Its Application to Geological and Geophysical Problems.* New York: John Wiley & Sons, Inc., xvi+399 pp.

KAY, MARSHALL, and COLBERT, EDWIN H. 1965. *Stratigraphy and Life History.* New York: John Wiley & Sons, Inc., vi+736 pp.

KURTÉN, BJORN. 1968. *Pleistocene Mammals of Europe.* Chicago: Aldine Publishing Co., viii+317 pp.

———. 1969. "Continental Drift and Evolution." *Scientific American,* Vol. 220, No. 3, pp. 54–64.

———. 1971. *The Age of Mammals.* New York: Columbia University Press, 250 pp.

LE PICHON, XAVIER. 1968. "Sea-Floor Spreading and Continental Drift." *Journal of Geophysical Research,* Vol. 73, pp. 3661–3697.

LEY, WILLY. 1969. *The Drifting Continents.* New York: Weybright and Talley, 90 pp.

MAYR, E., ed. 1952. "The Problem of Land Connections across the South Atlantic, with Special Reference to the Mesozoic." *Bulletin of the American Museum of Natural History,* Vol. 99, pp. 85–258.

NEUMAYR, M. 1887. *Erdgeschichte.* Leipzig: Bibliogr. Institut, Vol. 1, 653 pp., Vol. 2, 879 pp.

QUAM, LOUIS, ed. 1971. *Research in the Antarctic.* Washington: American Association for the Advancement of Science, Publ. No. 93, xv+768 pp.

RIDD, M. I. 1971. "South-East Asia as a Part of Gondwanaland." *Nature,* Vol. 234, pp. 531–533.

ROBINSON, PAMELA LAMPLUGH. 1967. "The Indian Gondwana Formations—a Review." *First Symposium on Gondwana Stratigraphy,* pp. 201–268.

ROMER, ALFRED SHERWOOD. 1966. *Vertebrate Paleontology.* 3rd ed. Chicago: University of Chicago Press, viii+468 pp.

SCHOPF, JAMES M. 1970. "Gondwana Paleobotany." *Antarctic Journal of the United States,* Vol. 5, pp. 62–66.

SCHUCHERT, CHARLES. 1932. "Gondwana Land Bridges." *Bulletin of the Geological Society of America,* Vol. 43, pp. 875–916.

SIMPSON, GEORGE GAYLORD. 1947. "Holarctic Mammalian Faunas and Continental Relationships during the Cenozoic." *Bulletin of the Geological Society of America,* Vol. 58, pp. 613–688.

———. 1953. *Evolution and Geography.* Condon Lectures, Oregon State System of Higher Education, 64 pp.

SMITH, A. GILBERT, and HALLAM, ANTHONY. 1970. "The Fit of the Southern Continents." *Nature,* Vol. 225, pp. 139–144.

STOKES, WILLIAM LEE. 1960. *Essentials of Earth History.* Englewood Cliffs, N.J., x+502 pp.

STRAHLER, ARTHUR N. 1971. *The Earth Sciences.* 2d ed. New York: Harper & Row, vii+824 pp.

TARLING, D. H., and TARLING, M. P. 1971. *Continental Drift. A Study of the Earth's Moving Surface.* London: G. Bell and Sons, Ltd., 112 pp.

TAYLOR, F. B. 1910. "Bearing of the Tertiary Mountain Belt on the Origin of the Earth's Plan." Bulletin of the Geological Society of America, Vol. 21, pp. 179–226.

VINE, F. J. 1966. "Spreading of the Ocean Floor: New Evidence." *Science,* Vol. 154, pp. 1405–1415.

WEGENER, ALFRED. 1966. *The Origin of Continents and Oceans.* Translated from the fourth edition by John Biram. New York: Dover Publications, Inc., ix+246 pp.

WILSON, J. TUZO. 1963. "Continental Drift." *Scientific American,* Vol. 208, pp. 86–100.

ZUMBERGE, JAMES H. 1958. *Elements of Geology.* New York: John Wiley & Sons, Inc., viii+382 pp.

A GLOSSARY OF PALEONTOLOGICAL NAMES AND TERMS

The scientist is often accused of indulging in jargon: a charge not uncommonly leveled at paleontologists because of the many names of extinct animals and plants that punctuate their writings. "Why," asks the plaintive reader, "cannot common names be applied to these long-vanished inhabitants of the earth?" It would be nice.

But there are no common names for the hosts of fossil organisms, and there is no point in trying to invent vernacular names for animals (or plants) never seen in life by modern men. This only leads to imprecise circumlocutions.

This glossary is presented as an attempted aid to the reader who encounters for the first time many of the extinct animals here described. It also includes a few other terms used by paleontologists and a very few vernacular names of modern animals, which may be unfamiliar to some.

Every known distinct organism, living or extinct, has been given a scientific name consisting of a generic designation, which expresses larger relationships (the genus), and a specific or trivial name, which is limited to one distinct form, known as a species. In this book names of organisms are limited largely to genera.

Aardvark. A modern African mammal, the only living representative of the Tubulidentata. It subsists upon termites.

Agriochoerus. A North American Oligocene artiodactyl, having four-toed, clawed feet and a long tail. The agriochoerids were related to the oreodonts, so typical of the North American Oligocene.

Agrosaurus. A Jurassic coelurosaurian dinosaur from Australia.

Allosaurus. A gigantic carnivorous dinosaur from the Upper Jurassic of North America. The skull was large, with bladelike teeth, the hind limbs were strong and birdlike, and the relatively small forelimbs had sharp claws.

Amphibia. The most primitive tetrapods, constituting a Class of the vertebrates. Modern amphibians undergo a metamorphosis during life from a fishlike tadpole to the adult stage. Extinct forms presumably also underwent this change.

Amygdalodon. A sauropod dinosaur from the Lower Jurassic of South America.

Anchitherium. A Miocene horse, widely spread through Holarctica. It had primitive, low-crowned cheek teeth, and three functional toes on each foot.

Ankylosauria. A suborder of ornithischian dinosaurs, the armored dinosaurs, of Cretaceous age.

Anthracotheres. Rather primitive and widely distributed artiodactyls, of Eocene to Pleistocene age. The skull was elongated, the feet were four-toed.

Apatosaurus. A gigantic sauropod dinosaur from the Upper Jurassic of North America.

Apidium. A primitive Oligocene monkey from North Africa.

Archaeopteryx. The first bird, from the Upper Jurassic of central Europe. The skeleton was reptilian and there were teeth in the jaws. The preserved specimens show clear imprints of feathers in the rock.

Archaeotherium. A large "piglike" artiodactyl from the Oligocene and Miocene of Eurasia and North America. The head was very large; there were two functional toes on the feet.

Arctostylops. A primitive North American notoungulate of Eocene age, related and possibly ancestral to the notoungulates or southern hoofed mammals that lived in South America during much of Tertiary time.

Arsinoitherium. A large "rhinoceroslike" Oligocene mammal from North Africa. This quite isolated genus was characterized by a pair of enormous transverse horns on the skull.

Artiodactyla. The even-toed hoofed mammals, consisting of various extinct groups and the modern pigs and peccaries, hippopotamuses, camels, tragulids, deer, giraffes, pronghorns, antelopes and cattle.

Astrapotherium. A large, Tertiary South American herbivorous mammal with short, broad feet and with rather large canine tusks. Various astrapotheres lived in South America during Paleocene through Miocene times.

Australian. One of the modern zoogeographic realms, comprising Australia, New Guinea and certain adjacent islands.

Australopithecus. An advanced "man-ape" from the Pleistocene of Africa.

Barosaurus. A gigantic sauropod dinosaur from the Upper Jurassic of North America.

Bassariscus. The modern "ring-tail" of the southwestern United States. A persistent Miocene-type of raccoon. Miocene fossils are known.

Bienotherium. A tritylodont, or advanced mammallike reptile, from the Upper Triassic of China.

Bison. The North American "buffalo" and the European Wisent, of Pleistocene and Recent age.

Borhyaena. A carnivorous marsupial from the Tertiary of South America. There were numerous genera of borhyaenids.

Borophagus. A large, heavy canid, or "dog," from the Pliocene of North America.

Bos. Pleistocene and Recent cattle.

Bothriodon. An anthracothere (artiodactyl) from the Oligocene of Europe and North America, and the Oligocene and Miocene of Africa. The skull was elongated.

Brachiosaurus. The largest of the sauropod dinosaurs, and the largest animal ever to live on the land. From the Upper Jurassic of North America and eastern Africa.

Cacops. A small labyrinthodont amphibian from the Lower Permian of North America.

Caenopus. A hornless rhinoceros from the Oligocene of North America.

Cambrian. The first geologic period of the Paleozoic era, dating from about 600 to 500 million years ago.

Camptosaurus. A primitive ornithopod dinosaur of small to medium size from the Upper Jurassic of North America.

Captorhinus. A small, primitive reptile from the Lower Permian of North America. The captorhinomorphs were typical of the Cotylosauria, the stem reptiles.

Carboniferous. The fifth period of the Paleozoic era, ranging in age from about 345 to 280 million years ago. The Carboniferous was a period of extensive coal swamps.

Carnivora. The order of meat-eating mammals, including the extinct creodonts and the viverrids or civets, hyaenas, cats, mustelids, dogs, raccoons, bears, and the aquatic sea-lions, seals and walruses.

Cenozoic. The last era of geologic history, beginning about 60 million years ago and extending to present time. The "Age of Mammals."

Ceratopsia. A suborder of ornithischian dinosaurs, the horned dinosaurs, from the Cretaceous of North America and northern Asia.

Ceratosaurus. A giant carnivorous dinosaur with a small horn on the front of the skull. From the Upper Jurassic of North America.

Cervus. The Pleistocene and Recent stag of Holarctica. The Wapiti or "elk" of North America.

Cetacea. The whales, porpoises and dolphins.

Cetiosaurus. A sauropod dinosaur from the Upper Jurassic of Europe.

Chalicotheres. Perissodactyls of Tertiary age from Eurasia and North America. The three-toed feet have claws instead of hooves.

Chasmatosaurus. A thecodont reptile from the Lower Triassic of South Africa. *See Proterosuchus*.

Chirotherium. Fossil footprints characteristic of the Lower Triassic of Europe and North America.

Cisticephalus. A small dicynodont (therapsid) reptile from the Upper Permian of South Africa.

Coelophysis. A slender, lightly-built coelurosaurian dinosaur from the Upper Triassic of North America.

Coelurosauria. Small theropod dinosaurs, with strong, birdlike hind limbs, small forelimbs with hands adapted for grasping, supple necks and slender skulls.

Colugo. The "flying lemur" of the Orient—a mammal with membranes between the limbs enabling it to glide. Order Dermoptera.

Condylarthra. Primitive hoofed mammals from the Paleocene and Eocene of Eurasia, North and South America.

Coryphodon. A rather large herbivorous mammal of the ancient order Amblypoda, from the Paleocene and Eocene of North America and the Eocene of Europe.

Corythosaurus. An Upper Cretaceous ornithopod dinosaur from North America. The jaws were broad, like a gigantic duck bill; there was a large, bony crest on the top of the skull.

Cotylosauria. An order of primitive reptiles, directly descended from certain labyrinthodont amphibians. The cotylosaurs were probably the stem reptiles, from which other reptilian orders evolved.

Creodonta. Archaic carnivorous mammals.

Cretaceous. The last of the three Mesozoic periods, ranging in age from about 130 to 65 million years ago.

Crocodilia. A reptilian order made up of varied extinct forms, and of the modern crocodiles, alligators and gavials.

Cynognathus. An advanced mammallike, or therapsid, reptile from the Upper Triassic of South Africa and South America. The skull was rather doglike in form.

Daedicurus. A giant armored glyptodont—an edentate related to the modern armadillo—from the Pliocene and Pleistocene of South America.

Daptocephalus. A very large dicynodont (therapsid) reptile from the Upper Permian of South Africa.

Dasypus. The Pliocene to Recent armadillo.

Devonian. The fourth Paleozoic period, ranging from about 395 to 345 million years ago. The first land-living vertebrates, the ancestral amphibians, arose in late Devonian time.

Diadectes. A Lower Permian amphibian (according to some authorities) or cotylosaurian reptile (according to others) from Texas.

Diadiaphorus. A South American hoofed mammal of Miocene-Pliocene age, very horselike in adaptations but quite unrelated to the horses. It is of the Order Litopterna.

Diceratherium. A small rhinoceros with a pair of transverse horns on the front of the skull. From the Miocene of North America.

Dicynodon. A Permo-Triassic therapsid reptile from Africa, Europe, Asia and South America. It had a distinctive skull with beaked jaws, and with a single pair of tusks in the skull.

Dicynodontia. The dicynodonts are found on all of the continents in Permian and/or Triassic sediments.

Didelphis. The American opossum.

Dimetrodon. A large pelycosaurian reptile characterized by a deep skull with large, bladelike teeth, and in life by a huge, longitudinal sail on the back. From the Lower Permian of North America.

Dinosaurs. The ruling reptiles of Mesozoic times. There were two orders of dinosaurs, the Saurischia and Ornithischia.

Dinotherium. A proboscidean mammal with large, recurved tusks in the lower jaw. From the Miocene and Pliocene of Eurasia, and the Miocene-Pleistocene of Africa.

Diplocaulus. A bizarre amphibian (Order Nectridia) from the Lower Permian of North America. The large skull was very flat and extremely broad, shaped like a broad arrowhead. The limbs were small.

Diplodocus. A long and slender sauropod dinosaur from the Upper Jurassic of North America.

Diprotodon. A very large Pleistocene marsupial found in Australia; something like a wombat but as large as a rhinoceros.

Diprotodonts. The suborder of marsupials containing the phalangers, kangaroos, wombats and various extinct forms.

Dryopithecus. A Miocene-Pliocene ape found in Europe and Africa.

Edaphosaurus. A pelycosaurian reptile from the Lower Permian of North America and Europe. There was a longitudinal sail down the back, supported by elongated vertebral spines, on which there were bony crossbars. The teeth were small and numerous.

Edentata. The order of toothless or peg-toothed mammals, containing the extinct glyptodonts and ground sloths and the living armadillos, tree sloths and ant-bears.

Elephas. A genus of elephants, containing various extinct forms and the modern Asiatic elephant.

Endothiodon. A dicynodont (therapsid) reptile from the Upper Permian of South Africa.

Entelodon. A large "piglike" artiodactyl from the Oligocene of Europe.

Eocene. The second epoch of the Tertiary period, extending from about 55 to 37 million years ago.

Eohippus. The earliest horse (more properly *Hyracotherium*) from the Eocene of North America and Europe.

Eosuchia. An order of reptiles, some of which were ancestral to the lizards. From the Permian and Triassic of Africa; but some few eosuchians persisted into later periods in other parts of the world.

Equus. The Pleistocene and Recent horses.

Eryops. A large, heavy labyrinthodont amphibian from the Lower Permian of North America.

Ethiopian. The zoogeographical realm including Africa south of the Atlas Mountains.

Fauna. A natural association of animals.

Flora. A natural association of plants.

Galechirus. A small pelycosaurian reptile from the Upper Permian of South Africa.

Glossopteris. A Permian plant, widely distributed in the southern hemisphere.

Glyptodon. A gigantic armored edentate, related to the modern armadillo, from the Pliocene and Pleistocene of South America.

Gomphodont. A term often applied to certain alvanced mammallike reptiles having broad cheek teeth.

Gorgonopsia. A subdivision of therapsids, or mammallike reptiles, from the Permian of Africa and Europe. Many gorgonopsians were characterized by large, saberlike canine teeth.

Hemicyon. A large doglike bear, perhaps close to the origin of the bears. From the Miocene of Eurasia and North America.

Hesperornis. A loonlike bird of Cretaceous age from North America.

Hipparion. An advanced horse, but with three toes retained on each foot, that arose in North America at the beginning of Pliocene times, and from there quickly spread through the Old World.

Hippidion. A Pleistocene South American horse, derived from Pliocene invaders from North America.

Holarctic. The zoogeographic realms embodying North America, and Eurasia north of the Himalaya Mountains.

Hoplophoneus. An Oligocene saber-toothed cat found in North America.

Hyaenodon. A large Eocene to Miocene creodont, or primitive carnivore, from North America, Eurasia and Africa.

Hypertragulids. Small North American deerlike artiodactyls, similar to the modern Oriental tragulids, of Eocene to Miocene age. A single ancestral genus of Eocene age is known in Mongolia.

Hypohippus. A Miocene-Pliocene horse of North America and Asia, retaining primitive characters, such as low-crowned teeth, and three functional toes on each foot.

Hypsilophodon. A small, primitive ornithopod dinosaur, from the Cretaceous of Europe.

Hyracodon. A lightly built, hornless rhinoceros from the Oligocene of North America.

Hyracoidea. The hyraxes or conies of Africa and the Middle East. Oligocene to Recent.

Hyracotherium. The earliest horse, often designated as *Eohippus*. A small perissodactyl with four toes on the front feet and three on the hind feet. Eocene of North America and Europe.

Ichthyosauria. Mesozoic reptiles of porpoiselike form, fully adapted for a marine life.

Ichthyostegalia. The first land-living vertebrates; the most primitive of the labyrinthodont amphibians. Especially well-known from the Upper Devonian of Greenland.

Ictidosauria. Advanced mammallike reptiles from the Triassic of Africa, Eurasia, North and South America and the Lower Jurassic of Europe. The tritylodonts, of almost world-wide distribution in the late Triassic, were ictidosaurs.

Iguanodon. A lower Cretaceous ornithopod dinosaur from Europe. Closely related forms are known from Asia, North America, Africa and Australia. The first dinosaur to be scientifically described.

Inostrancevia. A gigantic gorgonopsian (therapsid) saber-toothed reptile, from the Upper Permian of Russia.

Insectivora. The most primitive of the placental mammals, appearing in the Cretaceous and continuing into Recent times. Modern forms are shrews, moles and hedgehogs.

Jeholosauripus. Assigned to dinosaurian footprints from the Cretaceous of China.

Jurassic. The second of the three Mesozoic periods, ranging from about 190 to 130 million years ago.

Kannemeyeria. A large dicynodont (therapsid) reptile from the Lower Triassic of South Africa and South America.

Labyrinthodontia. The oldest and the dominant amphibians of late Devonian through Triassic times.

Lagomorpha. The hares, rabbits and pikas.

Laplatasaurus. A sauropod dinosaur from the Upper Cretaceous of South America, Madagascar and India.

Leptobos. An early type of cattle from the Pleistocene of Eurasia.

Liassic. A term often used for the Lower Jurassic.

Libytherium. A large giraffe from the Pliocene of Africa.

Litopterna. An order of South American hoofed mammals, living through the extent of Cenozoic time.

Lufengosaurus. A rather large prosauropod dinosaur from the Upper Triassic of China.

Lycaenops. A gorgonopsian (therapsid) reptile from the Upper Permian of South Africa.

Lystrosaurus. A highly specialized dicynodont (therapsid) reptile, found in the Lower Triassic of South Africa, Antarctica, India and China.

Macrauchenia. A South American hoofed mammal—a litoptern—of Pliocene and Pleistocene age.

Mammalia. Tetrapods which suckle the young on milk and which are characterized by hair and by internally-controlled body temperature.

Mammonteus. The woolly mammoth from the late Pleistocene of Eurasia and North America.

Mammoth. A general term to denote various Pleistocene elephants.

Marsupialia. Those mammals in which the young are born in a larval condition, and are commonly (but not invariably) carried by the mother in a pouch, or marsupium.

Mastodonts. Varied proboscideans, of Oligocene to Pleistocene age.

Megalosaurus. A large carnivorous dinosaur from the Jurassic and Lower Cretaceous of Europe and the Jurassic of Africa.

Megatherium. A gigantic ground sloth from the Upper Tertiary and Pleistocene of South America and the Pleistocene of North America.

Merycoidodon. An oreodont, a four-toed artiodactyl, abundant in the Oligocene of North America.

Mesohippus. A small, three-toed horse from the Oligocene of North America.

Mesosaurus. A small aquatic reptile of Permian age found only in South Africa and Brazil.

Mesozoic. One of the three great eras of earth history, characterized by an abundant fossil record. The Mesozoic Era, ranging from about 225 to 60 or 65 million years ago, was the time of reptilian dominance. It is often called the Age of Reptiles or the Age of Dinosaurs.

Metoposaurus. A large labyrinthodont amphibian from the Upper Triassic of North America, Europe and India.

Miocene. The fourth epoch of the Tertiary period, lasting from about 27 to 12 million years ago.

Moeritherium. The ancestral proboscidean from the Oligocene of North Africa. About the size of a large pig, with only moderately enlarged "tusks."

Moropus. A chalicothere, or clawed perissodactyl, from the Miocene of North America.

Multituberculata. An order of somewhat rodentlike Jurassic to Paleocene mammals, characterized by numerous cusps on the cheek teeth.

Mylodon. A large ground sloth from the Pleistocene of South and North America.

Nearctic. The zoogeographic sub-realm consisting of North America, including a part of northern Mexico.

Nectridia. An order of Upper Paleozoic amphibians.

Neotropical. The zoogeographic realm consisting of South America and much of Central America.

Notharctus. An Eocene lemur from North America.

Nothofagus. The modern southern-hemisphere beech.

Nothosauria. Marine reptiles of Triassic age.

Notoungulata. A large order of South American hoofed mammals of varied adaptations, living through the extent of Cenozoic time.

Oligocene. The third epoch of the Tertiary Period, lasting from about 37 to 27 million years ago.

Ophiacodon. A large pelycosaurian reptile from the Lower Permian of North America.

Ordovican. The second of the Paleozoic periods, ranging from about 500 to 440 million years ago.

Oreodonts. Eocene to Pliocene artiodactyls from North America. The oreodonts were vaguely similar to sheep in size, but their relationships were more with the camels.

Oriental. The zoogeographic realm composed of India south of the Himalayas, and southeastern Asia.

Ornithischia. One of the two orders of dinosaurs. In the Ornithischia the pubic bone was parallel to the ischium, giving the pelvis something of a birdlike form.

Ornitholestes. A very small theropod dinosaur from the Upper Jurassic of North America.

Ornithomimus. An Upper Cretaceous theropod dinosaur, about the size of an ostrich, from North America and Asia.

Ornithopoda. A suborder of ornithischian dinosaurs, containing the hypsilophodonts, iguanodonts and hadrosaurs, or "duck-billed" dinosaurs.

Oxyaena. A large creodont, or primitive carnivore, from the Paleocene and Eocene of North America.

Palaeomastodon. A primitive proboscidean from the Oligocene of North Africa. In *Palaeomastodon* there were two large tusks in the skull, and two in the lower jaw, and obviously a trunk.

Palaeoscincus. A large, armored dinosaur from the Upper Cretaceous of North America.

Palaeostylops. A Paleocene notoungulate (related to hoofed mammals in South America) from Mongolia.

Palearctic. The zoogeographic sub-realm consisting of Europe and Asia north of the Himalaya mountains.

Paleocene. The oldest Tertiary epoch, lasting from about 65 to 55 million years ago.

Paleozoic. The first of the three eras of geologic history with an abundant fossil record, lasting from about 600 to 225 million years ago.

Pantodonta. An order of Paleocene and Eocene hoofed mammals, most of them of large size and "clumsy" aspect. These mammals, often called amblypods, are from North America, Europe and Asia.

Paramys. A primitive, squirrellike rodent from the Paleocene-Eocene of North America and the Eocene of Europe.

Paranthropus. A Pleistocene "man-ape" from South Africa.

Parasaurolophus. A North American Upper Cretaceous hadrosaur, or "duck-billed" dinosaur, with an enormously elongated crest on the skull.

Pareiasauria. Very large cotylosaurs, or primitive reptiles, from the Upper Permian of South Africa and Russia.

Pareiasaurus. A South African pareiasaur.

Pascualgnathus. An advanced mammallike reptile (therapsid) from Argentina.

Patriofelis. A creodont, or primitive carnivore, from the Eocene of North America.

Pelycosauria. An order of Permo-Carboniferous reptiles directly antecedent to the mammallike reptiles, or therapsids. The pelycosaurs are especially abundant in the Lower Permian of North America, and occur also in Europe and Africa.

Pentaceratops. A large horned dinosaur from the Upper Cretaceous of North America and Asia.

Perissodactyla. The odd-toed hoofed mammals, consisting of the extinct titanotheres and clawed chalicotheres and the persisting tapirs, rhinoceroses and horses.

Permian. The last of the Paleozoic periods, ranging from about 280 to 225 million years ago.

Phenacodus. A primitive hoofed mammal—a condylarth—from the Paleocene and Eocene of North America and the Eocene of Europe.

Phiomia. A primitive mastodont from the Oligocene of North Africa, rather similar to *Palaeomastodon.*

Phytosaurs. Triassic thecodont reptiles very similar to crocodiles in appearance as a result of parallel evolution. From North America and Eurasia.

Phytosaurus. A phytosaur, abundant in North America and Europe.

Pithecanthropus. A primitive man from the Pleistocene of Asia. (Now commonly placed within the genus *Homo.*)

Placental Mammals. The Eutheria; the majority of mammals in which the young are nourished by the placenta and are born in a relatively advanced state of development.

Placerias. A very large dicynodont reptile from the Upper Triassic of North America.

Placodontia. Marine reptiles of Triassic age.

Plateosauravus. A prosauropod dinosaur from the Upper Triassic of Africa.

Plateosaurus. A prosauropod dinosaur from the Upper Triassic of Europe.

Pleistocene. The last geologic epoch, ranging from about 2 or 3 million years to perhaps 20 thousand years ago. In a sense, we are still in the Pleistocene. The great Ice Age.

Plesiosauria. Marine reptiles of Mesozoic age.

Pliocene. The last Tertiary epoch, immediately preceding the Pleistocene. Ranging from about 12 to about 2 or 3 million years ago.

Poebrotherium. A small North American camel of Oligocene age.

Precambrian. An inclusive term to designate all geologic time before the Cambrian, as well as rocks and the few fossils of that age. The bulk of geologic time.

Pricea. A lizardlike reptile from the Lower Triassic of South Africa.

Primates. An order of mammals including the lemurs, tarsiers, monkeys, apes and man.

Proboscidea. An order of mammals including the extinct moeritheres, mastodonts, mammoths and the two living elephants.

Procolophon. A small Lower Triassic reptile, a cotylosaur, from South Africa and Antarctica.

Procyonids. The raccoons, coatis, ringtails, pandas and their ancestors.

Prolacerta. A lizardlike reptile from the Lower Triassic of South Africa. *Prolacerta* and *Pricea* are probably near the ancestry of the lizards.

Propliopithecus. A monkey from the Oligocene of North Africa.

Prosauropods. Large, heavy, Triassic saurischian dinosaurs, probably close to the ancestry of the sauropod dinosaurs.

Proterosuchus. An elongated thecodont reptile with long jaws, from the Lower Triassic of South Africa. Frequently designated *Chasmatosaurus.*

Protoceras. A horned artiodactyl, a hypertragulid, with six horns on the skull. From the Oligocene of North America. There are various protoceratids.

Pteranodon. A huge flying reptile with a wing span of 25 feet from the Upper Cretaceous of North America and Europe.

Pterosauria. The flying reptiles, of Jurassic and Cretaceous age.

Pyrotherium. A large South American mammal of Oligocene age. It shows some adaptations (tusks, form of teeth, and an elongated nose) vaguely like those seen in proboscideans, but is quite unrelated to the latter.

Quaternary. The last period of the Cenozoic era. Essentially equivalent to the Pleistocene epoch.

Reptilia. Tetrapods in which reproduction is direct, from a protected egg or by live birth. In living forms the body is covered by scales and the body temperature varies more or less as does the temperature of the environment.

Rhoetosaurus. A sauropod dinosaur from the Lower Jurassic of Australia.

Rhynchocephalia. An order of reptiles, including the Triassic rhynchosaurs, and the modern tuatara of New Zealand.

Rhynchosauria. Large, beaked rhynchocephalians from the Triassic of South America, Africa and Eurasia.

Rodentia. The gnawing mammals: squirrels, gophers, porcupines, rats, mice and a host of others. The most numerous mammals.

Rubidgea. A very large, saber-toothed gorgonopsian (therapsid) reptile from the Upper Permian of South Africa.

Saurischia. One of the two orders of dinosaurs, characterized by a "normal" reptilian pelvis.

Saurolophus. A large ornithopod "duck-billed" dinosaur from the Upper Cretaceous of North America and Asia.

Sauropoda. Gigantic saurischian dinosaurs of Jurassic and Cretaceous age.

Scelidosaurus. A Lower Jurassic stegosaurian dinosaur from Europe.

Scymnognathus. A carnivorous mammallike reptile (therapsid) from the Upper Permian of Africa.

Seymouria. A labyrinthodont amphibian or a primitive cotylosaurian reptile (according to the bias of the authority concerned) from the Lower Permian of North America. Tetrapods of this type are called seymouriamorphs.

Silurian. The third Paleozoic period, extending from about 440 to 395 million years ago.

Sirenia. The sea-cows.

Smilodon. A large Pleistocene saber-toothed cat from North and South America.

Sphenodon. The tuatara, a lizardlike rhynchocephalian from New Zealand.

Spinosaurus. A large carnivorous dinosaur from the Upper Cretaceous of North Africa.

Stegosauria. A suborder of ornithischian dinosaurs of Jurassic and Lower Cretaceous age. The plated dinosaurs.

Stegosaurus. An Upper Jurassic plated dinosaur from North America.

Stereospondyli. Labyrinthodont amphibians of Triassic age.

Stereosternum. A name applied to the mesosaurs in Brazil. Probably a synonym of *Mesosaurus.*

Styracosaurus. A horned dinosaur distinguished by long spikes on the frill of the skull. From the Upper Cretaceous of North America and northeastern Asia.

Syntarsus. A small coelurosaurian dinosaur from the Upper Triassic of Rhodesia, Africa. Very closely related to *Coelophysis* of North America.

Synthetoceras. A hypertragulid artiodactyl from the Pliocene of North America.

Taeniodonta. An order of North American Paleocene-Eocene mammals.

Tapinocephalus. A large therapsid reptile from the Permian of South Africa.

Tarbosaurus. A gigantic carnivorous dinosaur from the Upper Cretaceous of Mongolia. Very close to *Tyrannosaurus* of North America.

Tertiary. One of the two Cenozoic periods.

Tetrapod. Literally a four-footed vertebrate. The primary land-living vertebrates-amphibians, reptiles, birds and mammals, as opposed to the primary aquatic vertebrates, the fishes.

Thecodontia. Essentially Triassic reptiles with two openings on each side of the skull behind the eye. Ancestral to the crocodilians, flying reptiles, birds and dinosaurs.

Therapsida. The mammallike reptiles.

Theriodontia. The most advanced of the therapsid reptiles.

Theropoda. A suborder of saurischian dinosaurs. These are the carnivorous dinosaurs.

Thoatherium. A small, hoofed litoptern mammal from the Miocene of South America, showing remarkable convergent adaptations to the true horses of this age.

Thrinaxodon. An advanced therapsid or mammallike reptile with carnivorous adaptations. From the Lower Triassic of South Africa and Antarctica.

Thylacinus. The Tasmanian "wolf"; a carnivorous, doglike marsupial, possibly extinct.

Thylacosmilus. A large, saber-toothed marsupial from the Pliocene of South America, showing adaptations convergent to the saber-toothed cats.

Ticinosuchus. A large, thecodont reptile from the Middle Triassic of Europe.

Tillodontia. A little-known order of North American Paleocene-Eocene mammals.

Titanosaurus. A sauropod dinosaur from the Cretaceous of South America, Asia and Europe.

Titanotheres. Very large perissodactyls, paralleling the rhinoceroses, from the Eocene and Oligocene of North America, Europe and Asia.

Toxodon. A large, vaguely rhinoceroslike notoungulate from the Pliocene and Pleistocene of South America.

Tragulids. Small, Old World artiodactyls of Eocene to Recent age. The Recent tragulids, or "mouse-deer," of the Orient are like miniature, hornless deer.

Triassic. The first of the three Mesozoic periods, ranging in age from about 225 to 190 million years ago.

Trilophodon. A Miocene-Pliocene mastodont with long tusks in the skull and lower jaw. From North America, Eurasia and Africa.

Tritylodon. An ictidosaur; an advanced mammallike reptile, showing certain rodentlike adaptations, from the Upper Triassic of South Africa and Argentina. Very closely related forms are known from China and North America.

Tuatara. *Sphenodon,* the single living rhynchocephalian reptile. From New Zealand.

Typotheres. Notoungulates from the Eocene to Pleistocene of South America, some of them showing rather rabbitlike adaptations.

Tyrannosaurus. The largest of the carnivorous theropod dinosaurs from the Upper Cretaceous of North America.

Uintatherium. A gigantic, herbivorous mammal, with six horns on the skull, from the Eocene of North America.

Ungulate. An animal with hooves. A very general term.

Ursus. The late Cenozoic to Recent bears of Eurasia and North America.

Viverrids. The Old World civets, ranging from Eocene to Recent times. Some recent civets are the genet, mongoose and binturong.

Zalambdalestes. A Cretaceous insectivore from Mongolia. One of the earliest known placental mammals.

ILLUSTRATIONS: SOURCES AND CREDITS

The following illustrations were drawn especially for this book by Nova Young, under the direction of E. H. Colbert.

FIGURES 2, 26, 27, 30, 31, 34, 35, 36, 40, 42, 43, 44, 48, 50, 51, 52, 56, 57, 58, 64, 66, 69, 70, 71, 75, 80, 81, 82, 86, 87, 94, 95.

FIGURE 2 is adapted from Hallam, Anthony, 1967. *Palaeogeography, Palaeoclimatology, Palaeoecology,* Vol. 3, pp. 201–241.

FIGURE 26 is adapted in part from Dietz, Robert S. and Walter P. Sproll, 1970. *Science,* Vol. 167, p. 1613.

FIGURE 27 is adapted from Zumberge, James H., 1958. *Elements of Geology.* New York: John Wiley & Sons, Inc. and from Martin Halpern, 1970.

FIGURE 30 is adapted from Du Toit, A. L., 1957. *Our Wandering Continents.* Edinburgh: Oliver and Boyd, Second Gondwana Symposium, (in Haughton, 1972), and from J. F. Lindsay, 1972.

FIGURES 34 and 57 are adapted in part from Bullard, Edward, 1969. *Scientific American,* Vol. 221, pp. 68, 74–75.

FIGURES 35, 50, 64 and 70 are adapted in part from Dietz, Robert S. and John C. Holden, 1970, *Scientific American,* Vol. 223, pp. 30–41, with modifications from David H. Elliot, 1970.

FIGURE 56 is adapted from Hayes, Dennis E. and Walter C. Pitman III, 1970. *Antarctic Journal,* Vol. 5, pp. 70–76, from Strahler, Arthur N., 1971. *The Earth Sciences,* p. 447, and from the National Geographic Society.

FIGURE 16 was drawn by Pamela Lungé, under the direction of E. H. Colbert.

The following illustrations are from photographs (taken by E. H. Colbert).
FIGURES 7, 8, 9, 10, 17, 19, 20, 21, 22, 28, 29, 39, 41, 46, 47, 53, 62, 63, 68, 84.

The following photographic illustrations are from sources, as indicated.
FIGURE 1. Courtesy of Robert S. Dietz.
FIGURE 5. Peter J. Barrett, Institute of Polar Studies, The Ohio State University.
FIGURE 6. Chester Tarka, American Museum of Natural History.
FIGURES 11, 13, 14. Official U.S. Navy photographs.
FIGURE 12. Marc Gaede, Museum of Northern Arizona. Model made by Margaret M. Colbert, under the direction of E. H. Colbert.
FIGURES 15, 99. Marc Gaede, Museum of Northern Arizona.
FIGURES 23, 25. David H. Elliot, The Ohio State University.
FIGURE 55. E. H. Colbert.
FIGURE 65. Gilbert F. Stucker. Courtesy of Mr. Stucker.
FIGURE 67. Werner Janensch. Courtesy of Dr. Janensch.
FIGURE 76. American Museum of Natural History.
FIGURE 78 (in part). American Museum of Natural History.

The following illustrations are taken from books and scientific papers, the authors and dates of which are listed below.
FIGURE 3. Du Toit, A. L. 1957. *Our Wandering Continents.* Edinburgh: Oliver and Boyd, Westport, Connecticut, Greenwood Press.
FIGURE 4. Wegener, Alfred. 1966. *The Origin of Continents and Oceans.* (Translated by John Biram.) New York: Dover Publications, Inc.
FIGURE 24. Brink, A. S. 1954. *Nat. Museum Bloemfontein,* Vol. 1.
FIGURE 32. Bonaparte, J. F. 1967. *Palaeontology,* Vol. 10, p. 556.
FIGURE 33. Restoration by Margaret M. Colbert in Colbert, E. H. 1965. *The Age of Reptiles.* New York: W. W. Norton and Co., Inc.
FIGURE 37. Jarvik, Erik. 1955. *The Scientific Monthly,* Vol. 80, p. 152.
FIGURE 38. Romer, A. S., and Price, L. I. 1940. Special Paper No. 28. *Geological Society of America.* Case, E. C. 1907. *Carnegie Institution of Washington,* Publication No. 55.
FIGURE 45. Gregory, W. K., and Camp, Charles. 1918. *Bulletin of the American Museum of Natural History,* Vol. 38, Art. 15.
FIGURE 49. Krebs, Bernard. 1963. *Schweizerische Palaontologische Abhandlungen,* Vol. 81.
FIGURES 54, 59, 79, 89. Colbert, Edwin H. 1969. *Evolution of the Vertebrates.* New York: John Wiley & Sons, Inc.
FIGURES 60, 61. von Huene, F. 1956. *Paläontologie und Phylogenie der Niederen Tetrapoden.* Jena, Gustav Fischer Verlag.

FIGURE 72. Marsh, O. C. 1880. *Odontornithes.* Geological Exploration of the Fortieth Parallel, Vol. 7.

FIGURES 73, 74. Simpson, G. G. 1928. *American Museum Novitates,* No. 329; 1937. *Bulletin of the American Museum of Natural History,* Vol. 73, Art. 8.

FIGURES 77 and 78 in part. Kielan-Jaworowska, Zofia. 1969. *Hunting for Dinosaurs.* Cambridge, Mass.: The MIT Press.

FIGURE 83. O'Harra, C. C. 1920. *South Dakota School of Mines,* Bulletin No. 13.

FIGURES 85, 88. Scott, W. B. 1937. *A History of Land Mammals in the Western Hemisphere.* New York: The Macmillan Company.

FIGURES 90, 96, 98. Osborn, H. F. 1915. *Men of the Old Stone Age.* New York: Charles Scribner's Sons.

FIGURES 91, 92. Osborn, H. F. 1942. *Proboscidea,* Vol. II. American Museum of Natural History.

FIGURE 93. Kurtén, Bjorn. 1968. *Pleistocene Mammals of Europe.* Chicago: Aldine Publishing Company.

FIGURE 97. Kay, Marshall, and Colbert, Edwin H. 1965. *Stratigraphy and Life History.* New York: John Wiley & Sons, Inc.

INDEX

ABOUT THE AUTHOR

Dr. Edwin H. Colbert is Curator of Vertebrate Paleontology at the Museum of Northern Arizona in Flagstaff, Curator Emeritus of Vertebrate Paleontology at the American Museum of Natural History in New York, and Professor Emeritus of the same subject at Columbia University. He has collected dinosaur and other fossils in many parts of the world, including Europe, the Middle East, India, Australia and South America as well as in North America. The discovery by Dr. Colbert and scientists under his direction of fossil remains of the reptile *Lystrosaurus* in Antarctica in 1969 made headlines throughout the world because of its important confirmation of the theory of continental drift. Dr. Colbert holds the Gold Medal of the American Museum of Natural History for Distinguished Achievement in Science and the Daniel Giraud Elliot Medal of the National Academy of Sciences. He is a member of the National Academy. His home is in Flagstaff, Arizona.